How Rude!

ANIMALS THAT BURP, TOOT, SPIT, AND SCREECH TO SURVIVE

Written by Chana Stiefel
Illustrated by Anna Louise Oliver

union square kids
NEW YORK

Contents

Excuse Me!

A nasty burp! A toxic sneeze! A knockout fart!

If *you* did these things, someone might say, ***"How rude!"***

But the critters in this book aren't *trying* to be obnoxious. Many of their "rude" behaviors are amazing tactics they've adapted to survive in nature.

Throughout this book you'll encounter gross habits, from snot-shooting to spitting to barfing. You'll meet worms, spiders, sharks, and other animals that do these seemingly nasty things for surprisingly good reasons. You'll also hear from scientists who are investigating what humans can learn from these bold behaviors. That knockout fart? It's actually a stinky trick to stun prey. How wild is that? So, the next time a boisterous beast lets one loose, maybe give them a respectful pass.

BuurrPP!

CHAPTER 1

Snot Shooters

Ah-ah-ah-chooo! Green gobs of sticky snot shoot from your nose. ***Gross!*** Yet boogers are more than just icky goo. Snot traps dirt, dust, and microscopic invaders like bacteria and viruses. It blocks them from infecting your airways and making you sick. Mucus (the polite word for *snot*) also contains antibodies and enzymes—protective proteins that recognize bodily invaders and kill them.

A sneeze is one of your body's ways of blasting out microbes. (That's why you need to cover your nose and mouth when you sneeze—so you don't infect others by shooting germs into the air.) In addition, mucus lining your nose, mouth, and intestines keeps your body running smoothly, like a well-oiled machine.

In the animal world, some critters use slimy mucus in gobsmacking ways to protect themselves. And that's nothing to sneeze at!

FOUL FACTS

- The sneeziest creature on Earth is the ***iguana***. Iguanas sneeze regularly to rid their bodies of excess salt, a by-product of their digestive system.
- By some estimates, human sneezes travel at about a hundred miles (161 km) per hour. A single sneeze can send a hundred thousand germs into the air. So sneeze into your elbow, please!

Snot Shield

NAME: **ORANGE CLOWNFISH**
SPECIES: *AMPHIPRION PERCULA*

SIZE:	4.3 inches (11 cm) long
RANGE:	Tropical regions of the Indian and Pacific Oceans
HABITAT:	Coral reefs
STATUS:	May be threatened due to climate change, habitat destruction, and the global marine aquarium trade (catching and selling fish for aquariums)

You might recognize these clownfish from the movies. What you might *not* know is that clownfish are stellar snot producers. They hang out among anemones, sea creatures that attach themselves to the seafloor and wave their tentacles to attract prey. The tentacles are lined with stinging capsules called *nematocysts.* When prey touch the tentacles, they shoot poison darts that kill the fish and drag them into the anemone's mouth.

Clownfish, however, have developed an ingenious way to protect themselves from the stinging tentacles. They produce sticky mucus that creates a shield against the poison. They also "boogie on down," doing a fishy dance as they rub different parts of their body against the tentacles until they become immune to their lethal stings.

In the process, the clownfish and anemone apparently swap bacteria-laden mucus so that the anemone recognizes the clownfish as a friend, not food. In exchange for protection from predators and scraps of leftover anemone food, the clownfish chases away other predators and removes harmful parasites from the anemone. This two-way relationship is called *symbiosis.*

FAB FACT

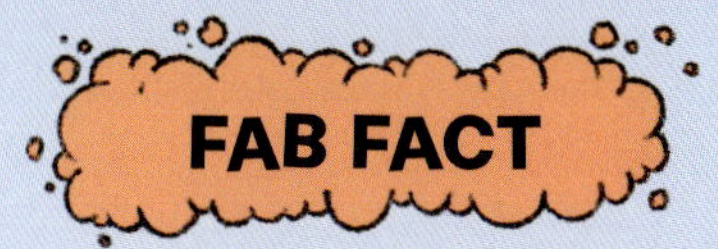

All clownfish are born male and have the ability to turn into females. Once they transform, they can't turn back.

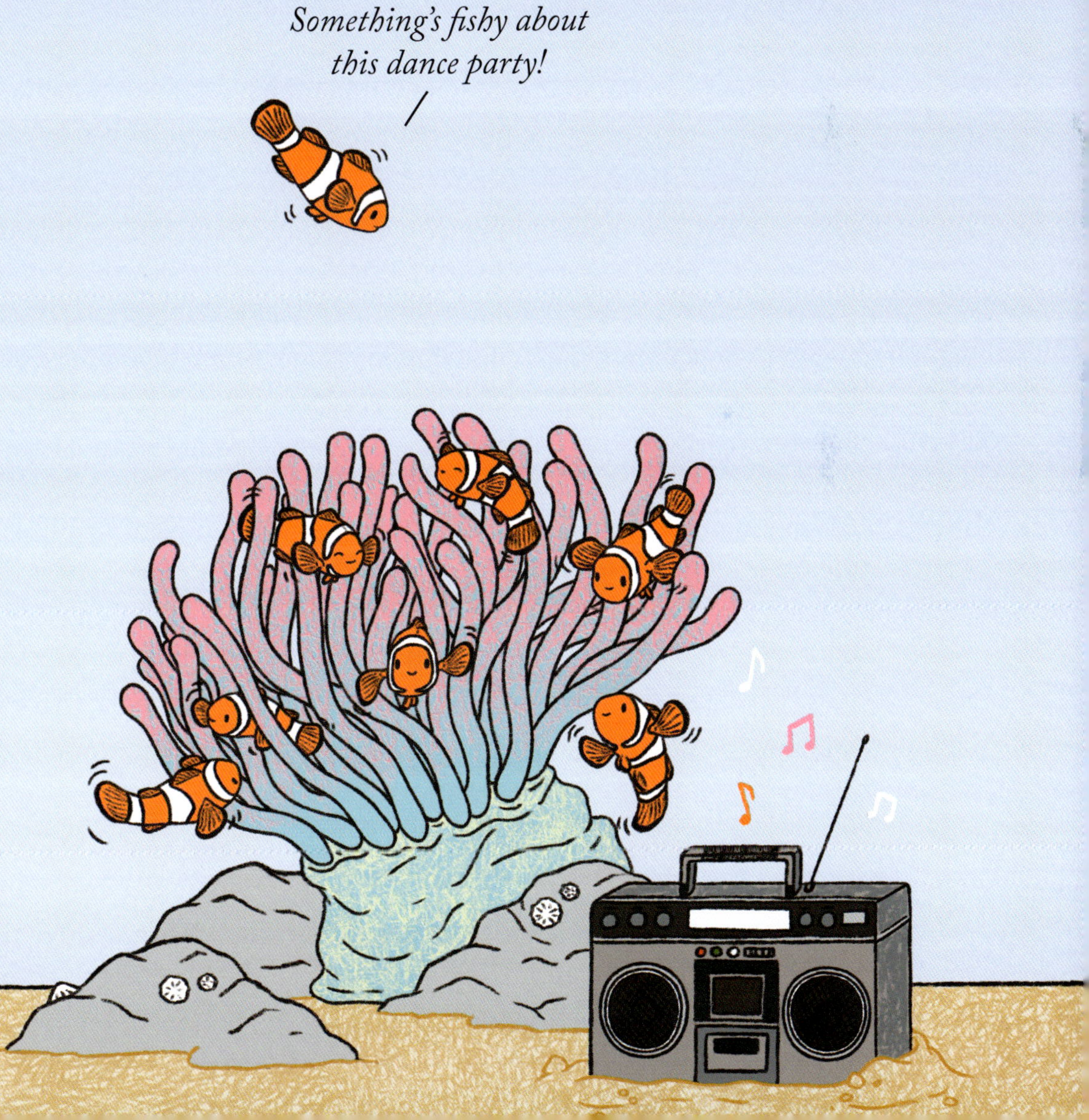

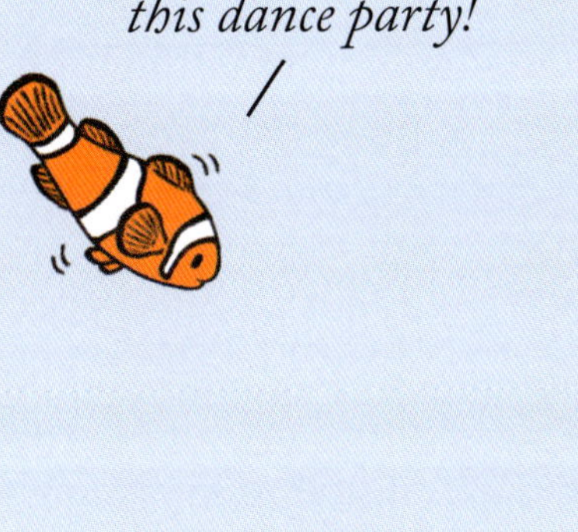

Mucus Maker

NAME: **BOOTLACE WORM**
SPECIES: *LINEUS LONGISSIMUS*

SIZE:	Typically 15 to 50 feet (5 to 15 m) but can grow up to 180 feet (55 m) long
RANGE:	Off the coast of Great Britain, Norway, Sweden, Denmark, and Iceland eastward to the Atlantic
HABITAT:	Beneath seafloor boulders and on muddy sand
STATUS:	Common

What could be stranger than a giant slimy worm? How about one that squirts **poisonous snot**? Sometimes growing up to 180 feet (55 m) long—half the length of a football field—the stretchy bootlace worm is one of the longest animals on Earth. It can even grow longer than a blue whale. Picture *that* slithering along on the seafloor! When threatened by predators—such as crustaceans, spineless creatures like crabs that have a hard exoskeleton—the worm releases gobs of sticky, toxic mucus. The substance, which gives off a stinky sewage smell, helps the worm slide through the mud and rocks to make a slippery getaway. What's more, the worm oozes the same mucus to trap its prey.

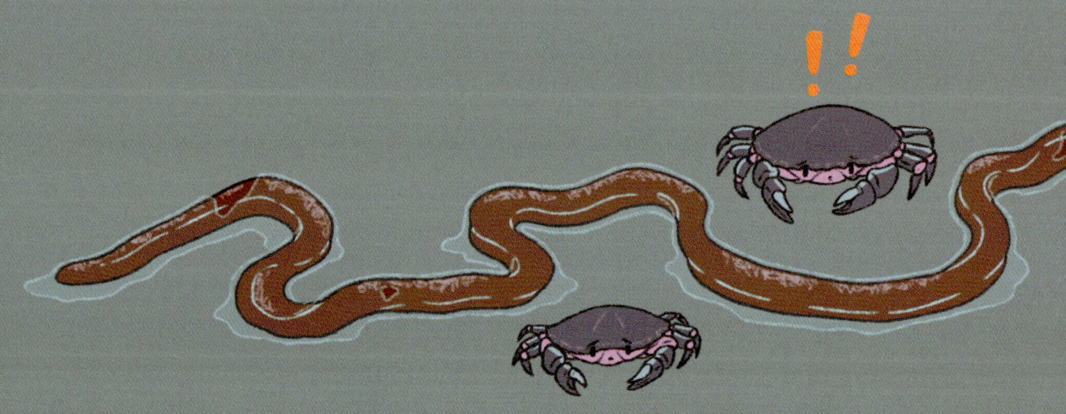

YUCKY OR USEFUL?

The toxin in bootlace-worm mucus can paralyze cockroaches. Scientists are currently exploring whether the poison can be used as a new insecticide.

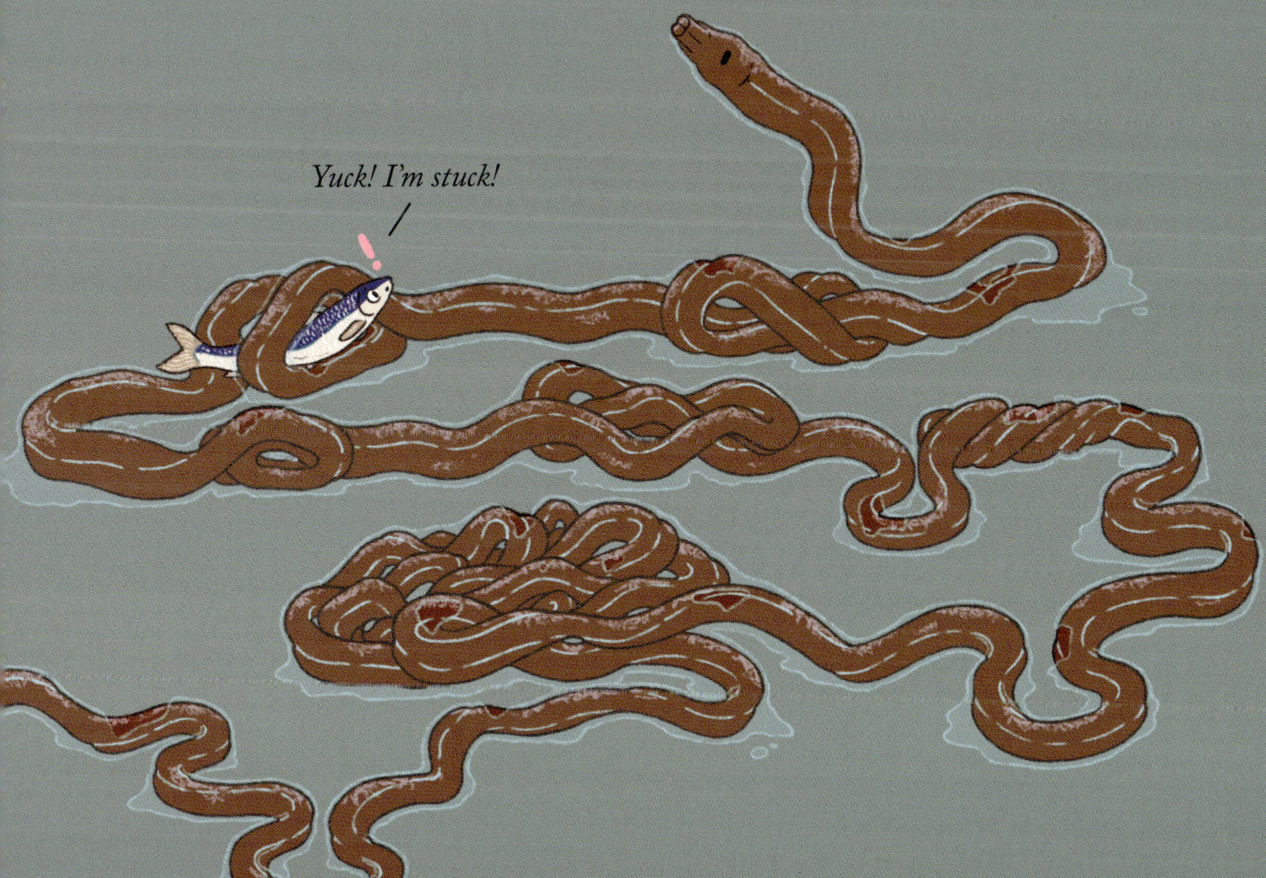

Super Slimer

NAME: **HAGFISH**
GENUS: *EPTATRETUS* (76 SPECIES)

SIZE:	Up to 4 feet, 2 inches (128 cm) long
RANGE:	Cold oceans around the world
HABITAT:	Muddy seafloor
STATUS:	At risk. In a study of 76 hagfish species, at least 12 percent were found to have a high risk of extinction.

Hagfish are spineless, jawless, and virtually blind eel-like creatures that wriggle along the seafloor. How do they defend themselves? With a ***slick trick***! When a predator attacks, hagfish ooze thick slime that clogs the gills and mouth of their attacker. Some predators survive the ordeal, but they learn to stay away.

In seawater, a small glob of the slime will expand to fill a bucket in less than half a second! To scrape the slime off its own body, the hagfish forms a knot and then slides through it from head to tail, wiping off the gelatinous goop. To prevent choking on its own slime, the hagfish sneezes out the slime through its nostril.

I would sneeze into my elbow if I had an elbow!

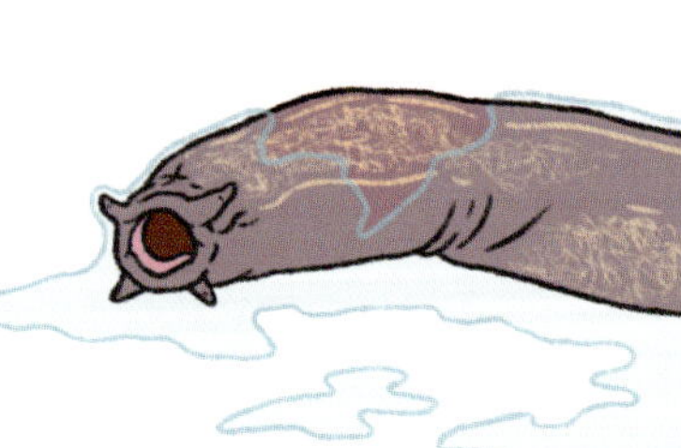

- Hagfish usually act as scavengers, feeding on decaying carcasses on the ocean floor. They have two rows of tooth-like structures that they use to chomp on dead sea creatures. This process helps keep their ocean environment clean and healthy.
- In 2017, ***a truck hauling hagfish*** overturned on an Oregon highway. The shipment was destined for South Korea, where the fish are a delicacy. Hagfish slime covered the roadway and at least one unfortunate snot-covered sedan!
- Hagfish Day is celebrated on the third Wednesday of October, a reminder to protect all ocean creatures, no matter how slimy or "rude."

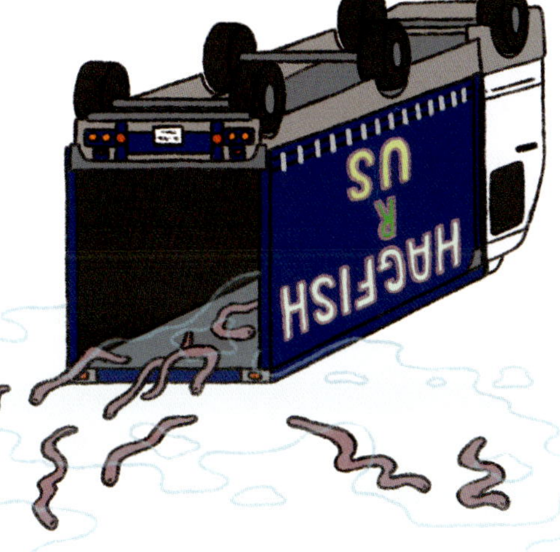

The protein threads produced in hagfish slime are a hundred times thinner than human hair and ten times stronger than nylon. Scientists are exploring uses for this tough material, including protective vests, food packaging, bungee cords, bandages, and airbags. Scientists are also studying hagfish slime to see if it can plug leaks in ships and oil rigs.

Slime Scientist

Antonio Cerullo, Biochemist

When did you first become interested in science?
My family owns a tree-service company, so I spent a lot of time thinking about plants. I realized I was one small part of something *really* big. I've always asked a lot of questions about living things. For example, if trees are solid wood, how does water travel from the ground to a tree's leaves?

What is your educational background?
I attended public schools and loved STEM and AP Biology (life science). In college, I majored in biology and then earned a Ph.D. in biochemistry (the chemical processes of living things) at the City University of New York (CUNY). I am the first in my family to complete college and graduate school.

How did you become interested in studying slime?
In my first year of graduate school, we rotated through various labs, which is like speed dating for nerds. I met a professor who said he was starting a new study of mucus. I thought that was super strange. But then I learned that mucus is magical! It's one of the most important materials on Earth. Nearly every animal on the planet produces mucus, yet no one fully knows what it's made of or understands how it works!

You helped create the world's first and largest mucus library. *Whaaat?*
The library is a super-cold freezer at CUNY that holds mucus samples from more than twenty-five different animal species. For some, we go on expeditions to collect the stuff ourselves. Other samples are provided by our partners from almost every continent who support our goal of studying every type of mucus on Earth. Scientists from various fields can access the mucus samples to study and share their data.

What is the basic "recipe" for mucus?
Mucus is mostly water, a bit of protein, some sugar, and salt. But every animal is unique, and their mucus is even more special. For example, a garden snail produces three different types of mucus, which have over seventy different proteins.

How do you collect mucus?
It depends on the animal. For the garden snail, we gently scoop the mucus oozing from its

back into a tube. (That mucus is a barrier that keeps the snail hydrated and protects it from germs.) Then we let the snail glide around a glass petri dish and we collect the "snail trail" from its foot. Finally, we hold a dish upside down, and the snail sticks to it by producing a gluey mucus from its foot.

How is animal mucus both yucky and useful to humans?

Because of its hydrating and wrinkle-reducing properties, snail mucus is used in cosmetics, such as moisturizers and face masks. It's part of a billion-dollar industry! Other researchers are studying hippo mucus, which coats the animals' skin in the hot African sun, as a possible ultraviolet (UV) shield and defogger for astronauts' visors.

What's the "rudest" animal behavior you've encountered (other than by humans)?

Off the Florida coast, upside-down jellyfish shoot blobs of mucus containing stingers, called *nematocysts*. While this may be an ingenious way of stunning prey, many divers emerge from the water covered in a painful rash from the stings, even if they never saw a jellyfish! This mysterious phenomenon is called "stinging water."

CHAPTER 2

Gas Passers

Burrrp! Toot, toot! Pardon me! Let's face it: Everybody burps and farts. Healthy people pass gas twelve to twenty-five times a day, sometimes more. When you eat or drink, you swallow air, which contains gases like nitrogen and oxygen. These gases pass through your digestive system when your body breaks down food. Some air escapes through your mouth (a burp), and some passes through your rear end (a fart, or *flatus*).

Another source of gas: Bacteria in your gut break down foods that your body doesn't use, and in the process, they produce gas. When a combo of gases like hydrogen, carbon dioxide, and methane mix with hydrogen sulfide and ammonia—***P.U.!*** You get a stinky fart.

Lots of animals pass gas, too. Some belch and some fart. But a variety of critters win the prize for Graceless Gas Passers. They cut the cheese to survive!

- ***Birds don't fart!*** Not even large ostriches. They don't have the same gas-producing bacteria in their guts as farting animals do.
- ***Zebras*** fart when startled.
- The proper scientific word for passing gas is *flatulence,* the ejection of gas produced during digestion in the stomach or intestines. The cruder word *fart* dates back to the fourteenth century. A fart's stench depends on what the animal ate, its health, and the bacteria in its gut.

It wasn't me! Really!

Belching Bovines

NAME: **COW**
SPECIES: *BOS TAURUS*

SIZE:	Varies by breed, ranges from 300 to 3,000 pounds (136 to 1,360 kg)
RANGE:	All continents except Antarctica
HABITAT:	Domesticated cows are found in agricultural areas (farms)
STATUS:	Common

Imagine if your burps were so strong they heated the planet! Cows, sheep, goats, buffalo, and camels all belch methane, a *greenhouse gas* that traps heat in the atmosphere and contributes to climate change. In fact, burping livestock account for 14 to 26 percent of the world's total greenhouse emissions.

Cows, along with these other grass-eating animals, are called *ruminants*. They have a specialized digestive system that allows them to digest grass and other plants. Their burps come from their *rumen,* one of four sections in their stomach. Inside the rumen, billions of microbes break down tough, high-fiber food like grass. In the process, the microbes give off hydrogen and carbon dioxide gas. More microbes, called *methanogens*, combine these gases to form methane. ***Buuurrrrp!***

You can't expect a cow to mind its manners. One solution for reducing methane in the atmosphere is for people to cut back on meat and dairy products. (Less demand for hamburgers and cheese means fewer belching cows.) Some scientists are experimenting with adjusting the diets of cattle to make them burp less.

FOUL FACT

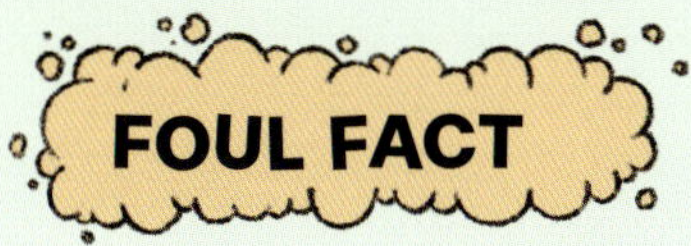

A cow burps between four and eleven cubic feet (120 to 320 liters) of methane per day. Methane is twenty-eight times more potent than carbon dioxide in trapping heat in the atmosphere.

YUCKY OR USEFUL?

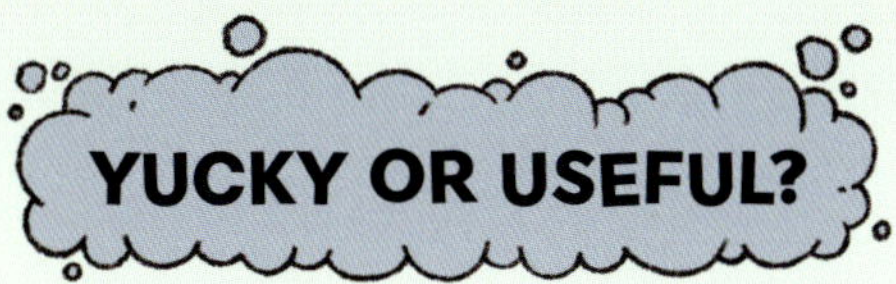

Scientists in Argentina have developed burp packs for cows that trap the methane they belch and turn it into clean energy. A tube is inserted into the cow's digestive tract to collect methane. The amount of natural gas collected in a single burp pack is enough to fuel a car for one day.

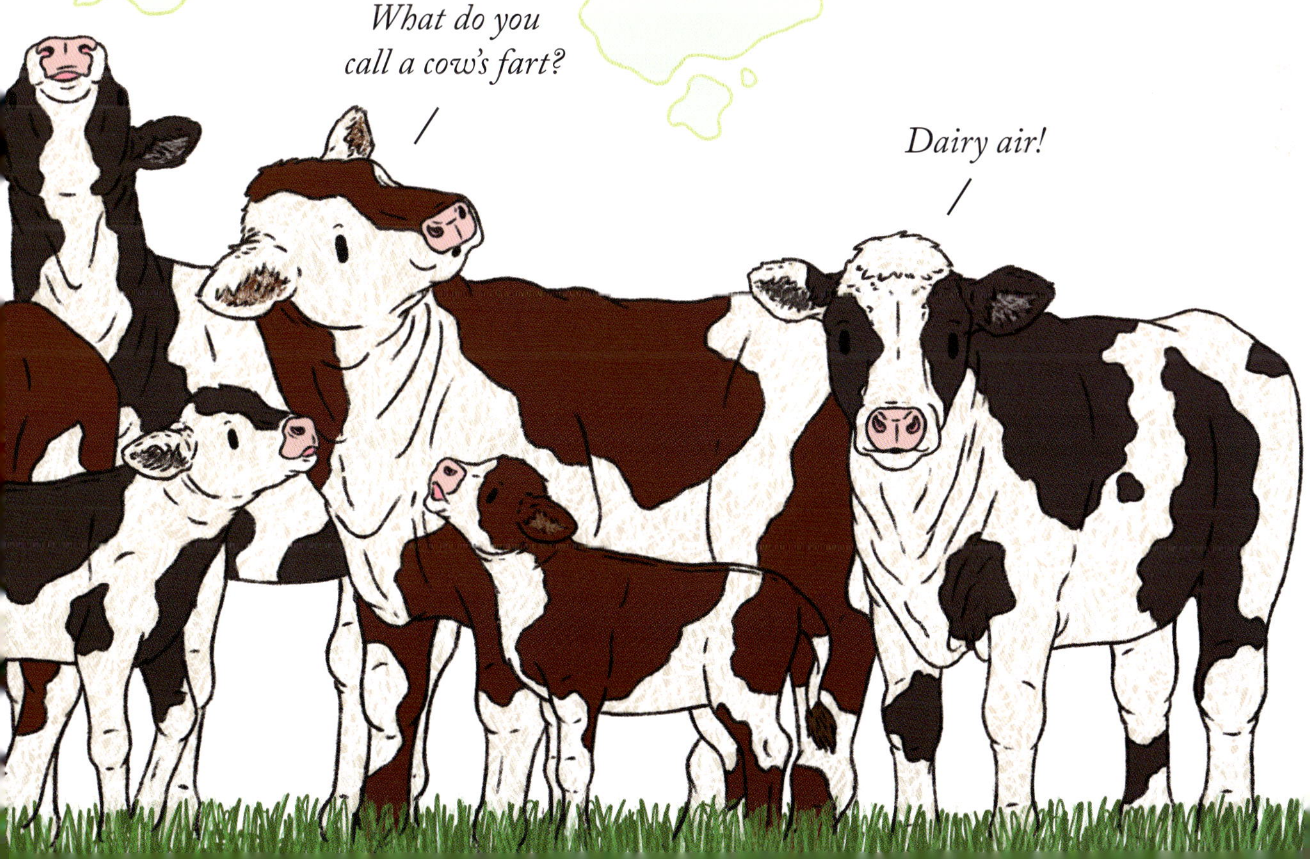
What do you call a cow's fart?
Dairy air!

School of Farts

NAME: **PACIFIC HERRING**
SPECIES: *CLUPEA PALLASII*

SIZE:	Up to 18 inches (46 cm) long
RANGE:	Off the coasts of Alaska and West Coast of the United States, Northeast Asia
HABITAT:	Coastal waters, surface to 1,300 feet (396 m)
STATUS:	Not threatened

Imagine if kids in your class passed gas to ***secretly communicate*** in code. Far(t)-fetched? Not if you're in a school . . . of herring! As night falls, herring gather in groups called *protective shoals*. How do they communicate when and where to meet up? Farting might be the key! Turns out that herring gulp air at the surface and then shoot air bubbles from their rear ends, emitting high-frequency "raspberry" sounds. Scientists call the sounds Fast Repetitive Tics, or FRTs (get it?!). Herring are known to have better hearing than other types of fish, so they may be letting off these FRTs to secretly communicate.

- Herring farts once sent the Swedish Navy scrambling. They thought the high-frequency FRTs were coming from Russian submarines.
- The scientists who discovered herring FRTs won the Ig Nobel Prize, which celebrates research so surprising that it makes people laugh, then think.

Phht! Phht!
Over here!

Gassy Sea Cow

NAME: **WEST INDIAN MANATEE**
SPECIES: *TRICHECHUS MANATUS*

SIZE:	Up to 14 feet (4.3 m) long; can weigh as much as 3,000 pounds (1,360 kg)
RANGE:	U.S. Atlantic and Gulf of Mexico coasts, throughout the Caribbean and as far south as Brazil's Atlantic coastline
HABITAT:	Shallow rivers, canals, saltwater bays, estuaries, and coastal areas
STATUS:	Threatened

West Indian manatees, also known as sea cows, are gentle mammals that graze on sea plants. A one-thousand-pound (453 kg) manatee can eat between 100 and 150 pounds (45 to 68 kg) of vegetation a day. That produces major gas in their gut. But manatees don't just let loose randomly. They have pouches in their intestines that store the gas. Why? These mammoth creatures hold in their farts to ***stay afloat***! When they pass gas, they sink. Maybe try that in the bathtub!

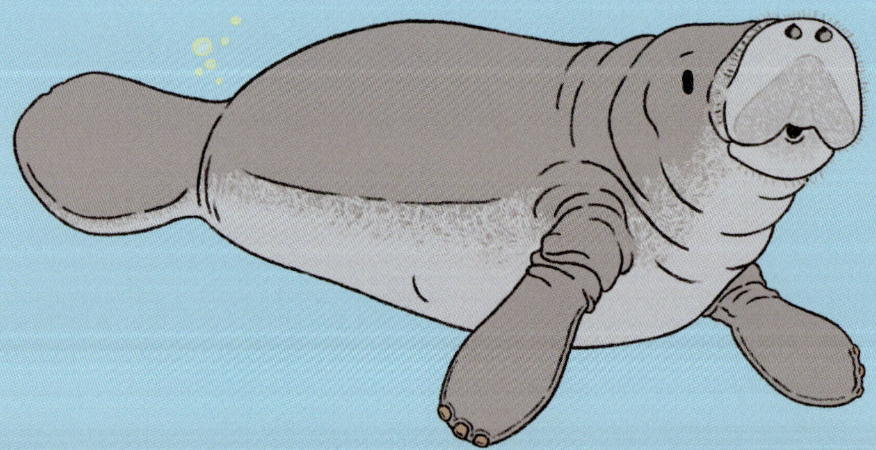

FOUL FACT

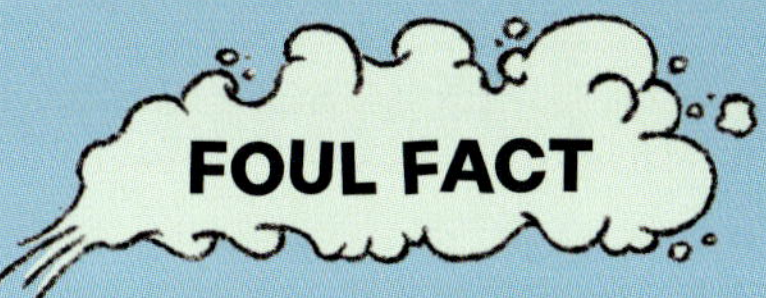

Manatees may have inspired the myth of mermaids. But perhaps not farting mermaids.

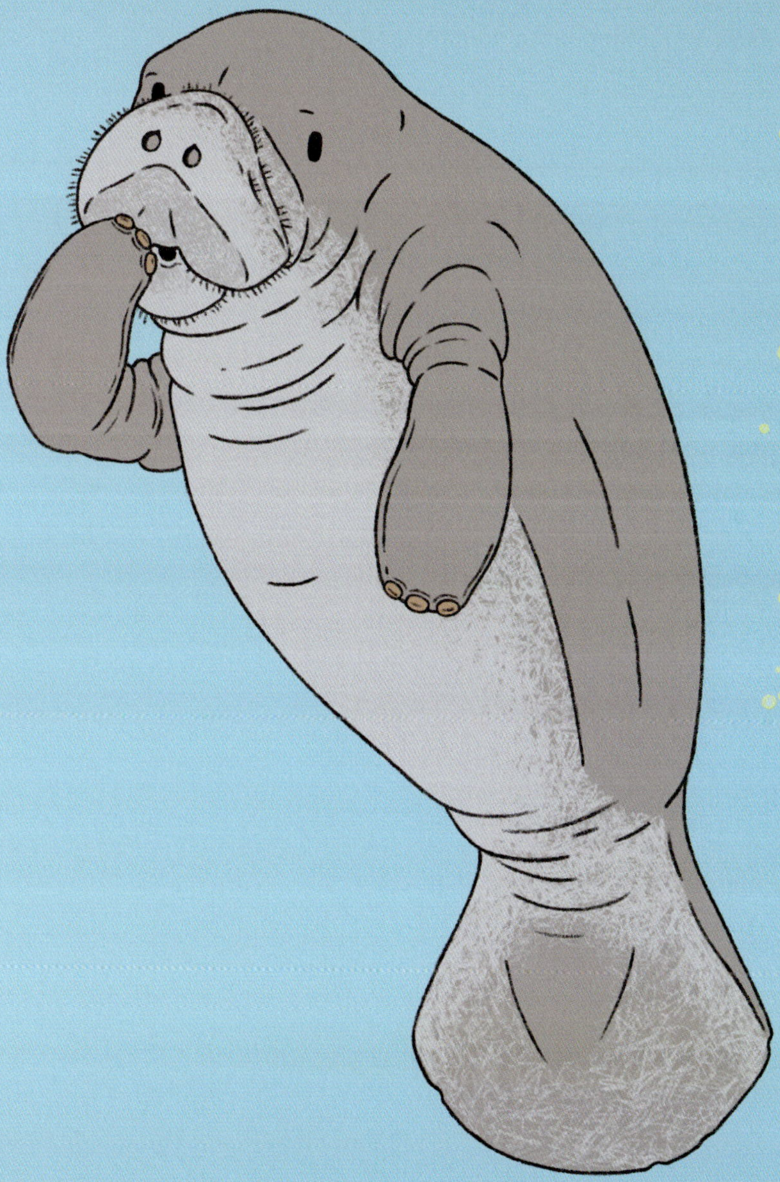

Silent but Deadly

NAME: **GREEN LACEWING**
SPECIES: *CHRYSOPERLA RUFILABRIS*

SIZE:	Adults are about 0.5 inches (13 mm) from head to wing tips and about 0.4 inches (10 mm) in body length
RANGE:	North America
HABITAT:	Field and tree crops, gardens, and in the wild
STATUS:	Common

The green lacewing is a delicate winged insect that sometimes lays eggs near termite nests. But termites, beware! When hungry lacewing larvae hatch, they wriggle into the nest, raise their bums, and let off a toxic toot. A chemical in the blast stuns the termites, paralyzing them for up to three hours. Meanwhile the larvae munch away on their termite lunch. Termites that aren't eaten die anyway. Now that's some ferocious flatulence!

Scientists have tested the deadly lacewing larvae farts on other types of insects, such as fruit flies, wasps, and booklice. But they were fart-resistant. No one knows why.

FOUL FACT

A deadly fart from a single lacewing larva can knock out six termites.

Boiling Beetle Juice

NAME: **BOMBARDIER BEETLE**
SPECIES: *BRACHINUS ELONGATULUS*

SIZE:	0.5 inches (1.27 cm) long
RANGE:	Central America, North America
HABITAT:	Moist areas under rocks, crevices, leaf litter
STATUS:	Common

Phht! Phht! Phht!

Excuse me! Excuse me! Excuse me!

Watch out, predators! When threatened, the bombardier beetle shoots a jet of boiling-hot chemicals from its rear end. How does this insect weaponize without harming itself? Scientists have studied this phenomenon using a high-powered X-ray machine called a *synchrotron*. What they found: Two chemicals combine to form a toxic liquid in a protective chamber inside the beetle. The chemical reaction produces intense heat and pressure. An explosive liquid shoots out of the beetle's behind along with—***poof!***—a dramatic puff of steam.

Some bombardier beetles spray in every direction, like a poisonous lawn sprinkler. Scientists have observed these insects pulsing three hundred to a thousand jets of gas each second, at speeds higher than 22 miles (35 km) per hour. That's enough fart-fire to chase away even the most gutsy predators.

Only a few attackers have developed ways to avoid the blast of hot beetle spray. Blue jays and some other birds rub the beetle's belly against their feathers or the sand. That protects them until the insect runs out of steam. Certain spiders wrap the beetle in silk to block their fiery farts.

Explosive Insect Investigator

Wah-Keat ("Wah-KET") Lee, Physicist

What sparked your interest in science?

I've always been interested in science. I remember going to a neighbor's house to watch the moon landing on TV as a child in 1969.

What is your educational background?

In college, I studied engineering and math. In graduate school, I earned a Ph.D. in physics, the study of matter and energy.

You work as a physicist . . . but you also study bugs? Explain!

I work as a physicist at a massive facility called a *synchrotron*. It's about the size of a football field. A synchrotron produces X-rays that are about a billion times more powerful than the X-rays at your dentist. Mostly we use the machine to study materials like batteries or metals on a super-microscopic level.

In 1999, I suggested to a colleague that we use the synchrotron to study an insect. He found an ant in a cobweb. While we were studying the ant under the powerful X-rays, it moved its leg. The machine produced images that I was certain no one had ever seen before. And something clicked! What if we use the synchrotron to study how insects function—for example, how they eat, breathe, and defend themselves?

So you bombarded the bombardier beetle with X-rays?

Yes, we collaborated with other scientists who study biology. We wanted to understand how the bombardier beetle sprays a boiling-hot toxin to defend itself when predators get close. First, we put the beetles in a fridge for a while to get them to "chill out." Then, one by one, we mounted the beetles in the machine. We had to wait for a long time to get one to spray. We tried dozens of times. We prodded them with a needle. Finally, one of the beetles let off a blast of toxic spray. We captured amazing images! We can do this because the synchrotron produces such bright X-ray pulses that we can capture images at extremely high speeds.

What did you learn?

It was incredible to see tiny chambers inside the beetle, where chemicals are produced. Then the chemicals combine, causing an explosion. It's like a kilogram of TNT exploding inside a human. Yet the beetle remains unharmed.

What are the toxic takeaways?

The images revealed that the pulses are controlled by the flexible walls of the reaction chamber and a valve that opens and closes, allowing a drop of chemicals to enter the chamber and set the explosion in motion. Studying the beetle's mechanisms might help scientists develop better blast-protection or lightweight propulsion systems.

What are you studying now?

I work on the synchrotron at Brookhaven National Laboratory in New York that uses X-rays to capture 3D details of things like corrosion forming on a metal surface, or rechargeable batteries breaking down. These images may help scientists make better materials.

Do you have a pet?

Yes, I have a loud dog named Sadie. We adopted her from an animal shelter.

What's the "rudest" animal behavior you've seen?

Sometimes dogs eat their own poop.

CHAPTER 3

Spitters

Pitooey! Spitting in public is not a favorite move for anyone nearby. It can spread harmful germs. But *saliva* (the scientific word for *spit*) is necessary for survival. Produced by glands in your mouth, saliva is mostly made of water mixed with a few other chemicals. It protects against harmful microbes, helps break down food to make it easier to swallow, starts the digestion process, and keeps your teeth and gums healthy (though you still have to brush!). Rude as it may seem, some animals use spitting as a strategy to help them navigate some sticky situations.

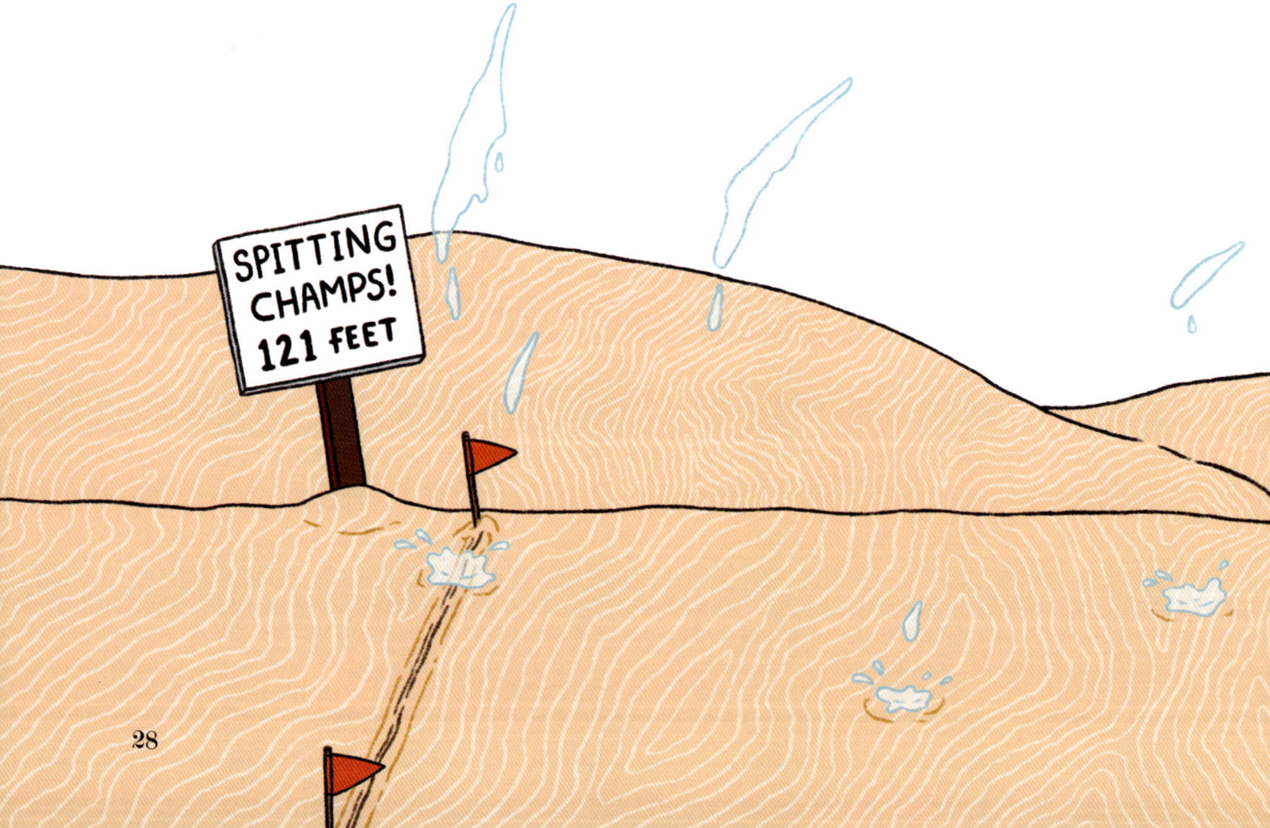

FOUL FACTS

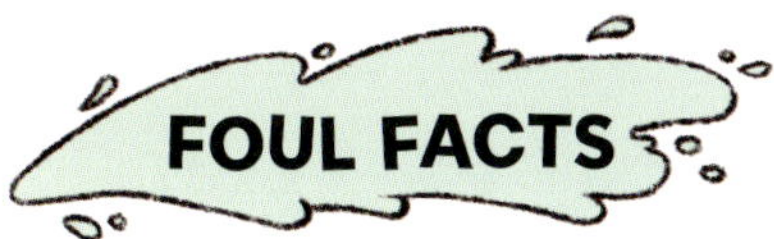

- The average person produces between 0.5 quarts and 1.6 quarts (0.5 and 1.5 liters) of spit per day, enough to fill a water bottle.
- When a ***camel*** feels threatened, its cheeks fill up with spit mixed with the contents of its stomach. Look out! ***Splork!*** Camels spit to surprise, distract, or scare away the source of their stress.
- The longest recorded distance that a camel has spit is 121 feet.

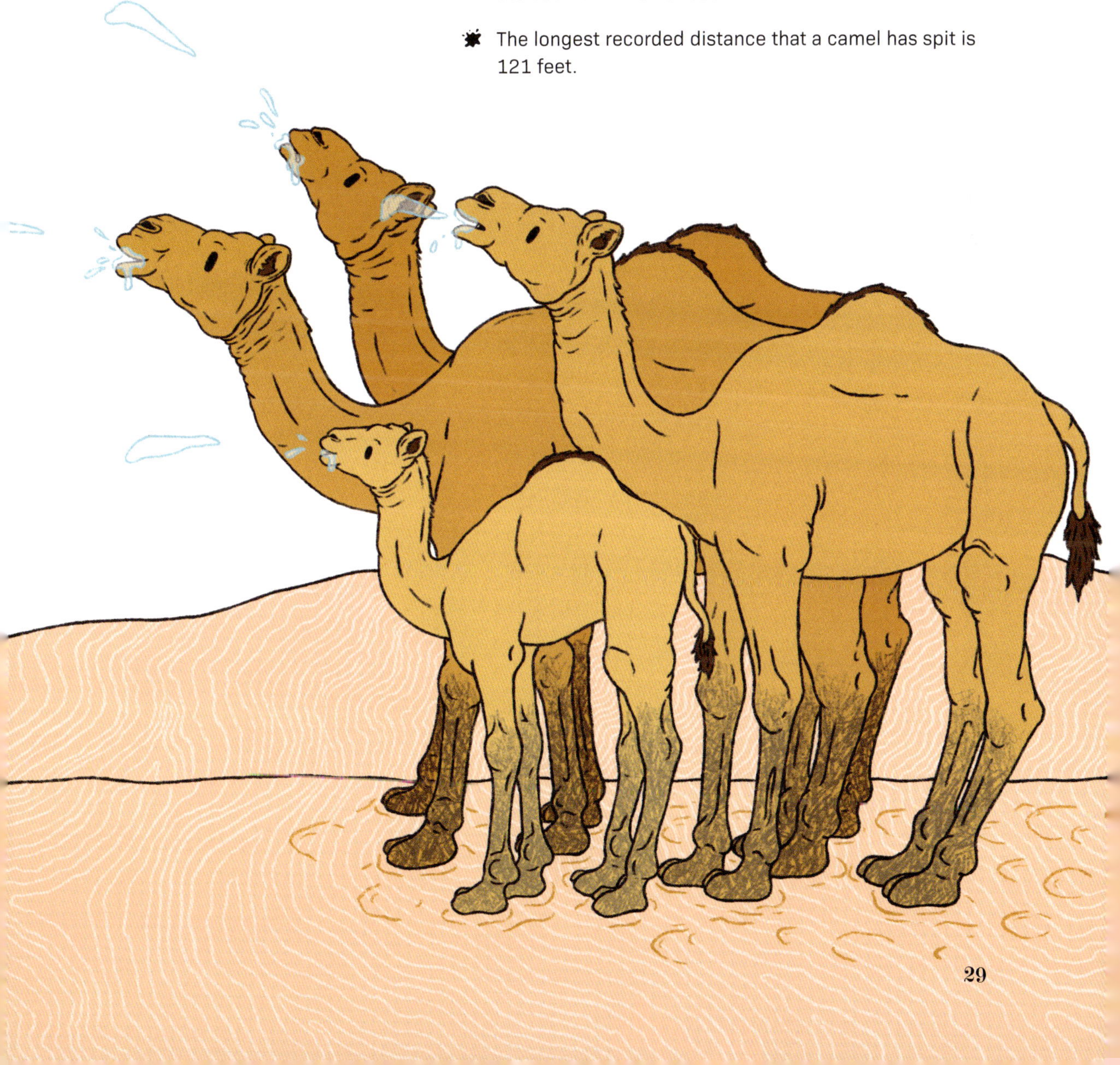

Sneaky Spritzer

NAME: **SPITTING SPIDER**
SPECIES: *SCYTODES THORACICA*

SIZE:	0.12 to 0.24 inches (3 to 6 mm) long
RANGE:	Eastern U.S., Britain, Sweden, and other European countries
HABITAT:	Temperate forests; when living near humans they prefer dark corners, cellars, and closets
STATUS:	Common

Some spiders spin attractive webs. Then they wait patiently to snare an unwitting fly. Not spitting spiders. In the dark of night, they prowl their territory looking for prey. When they spot a potential meal, ***SPITOOEY!*** They fire sticky silk to pin down their victim. Each strand of silk is coated with venom. Their poisonous spit can travel at almost ninety-eight feet (thirty meters) per second. It forms a superstrong net that traps the prey. If the prey tries to flee, the spider rolls it up like a bug in a rug. With a vampire-like bite, the spider injects the prey with paralyzing venom and deadly saliva that liquefies its insides. Then the spider slurps up its fresh catch.

Spitting spiders tend to hang out in dark corners or warm, cozy closets. Sometimes they spit to ward off predators. But don't worry! They're not harmful to humans.

Scientists are researching the proteins in spider venom to create new medicines to treat chronic pain, inflammation, and heart disease.

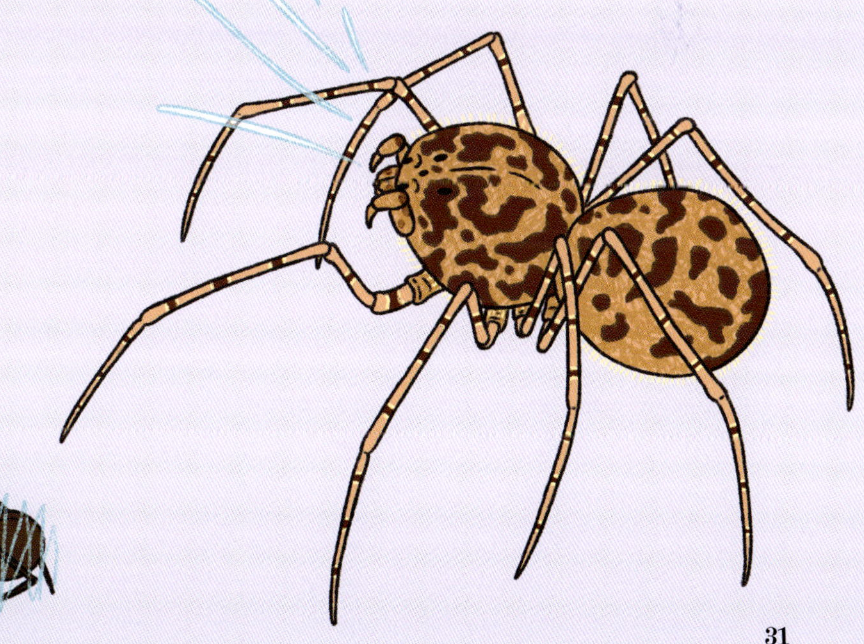

Green Grenade

NAME: **LLAMA**
SPECIES: *LLAMA GLAMA*

SIZE:	Adults are about 6 feet tall (1.8 m); 290 to 440 pounds (132 to 200 kg)
RANGE:	South America
HABITAT:	Natural habitat is high plateau covered with shrubs, stunted trees, and grasses at elevations ranging from 7,550 to 13,120 feet (2,300 to 4,000 m); domesticated commercial herds are found on farms in North America, Europe, and Australia
STATUS:	Common

Most llamas are sweet and gentle. And then there are the spitters. When annoyed, some llamas spit a slimy green gob of partially digested food a distance of about 15 feet (4.6 meters). They sometimes spit at other llamas to show dominance, an animal's way of showing who's boss. An adult llama might spit at a youngster who tries to eat its food. A female may shoot a green spit grenade at a flirting male to express her lack of interest. And a well-aimed ***pitooey*** can chase away a predator.

Most llamas don't spit at humans. But know the warning signs. Before coughing up a loogie, a llama flattens its ears against its head, looks you in the eyes, raises its chin, and begins to gurgle. If that happens, act cool. Avoid eye contact. And if you do get slimed, ***gahhhhhh!*** It's not harmful. Just time for a shower.

YUCKY OR USEFUL?

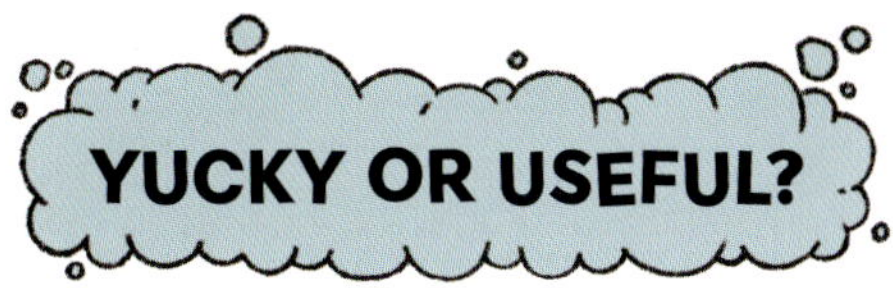

Researchers have discovered that tiny "super-immunity" molecules called *nanobodies*, found in the blood cells of llamas, could help protect against diseases like COVID-19.

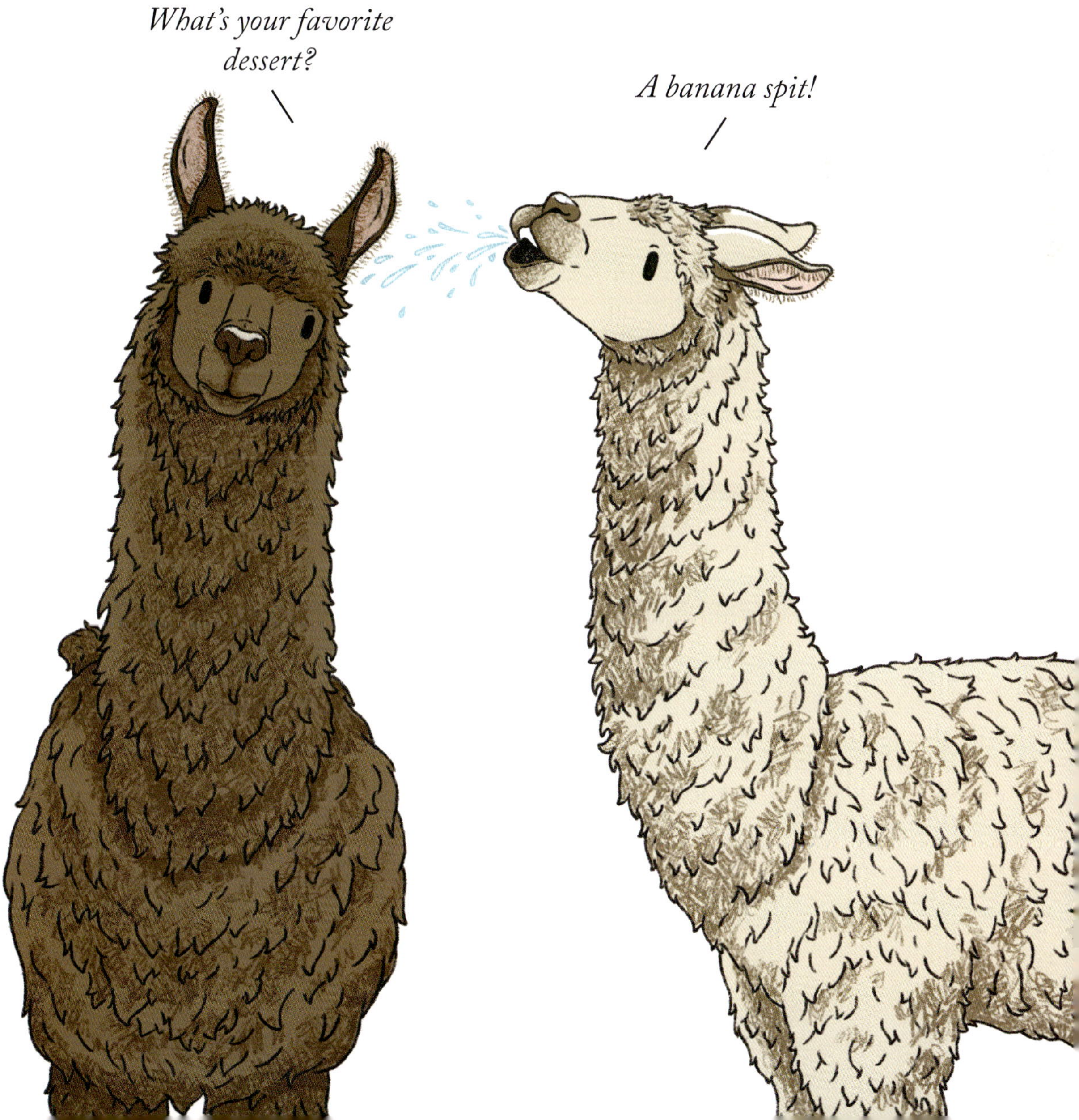

What's your favorite dessert?
A banana spit!

Fish Spit

NAME: **BANDED ARCHERFISH**
SPECIES: *TOXOTES JACULATRIX*

SIZE:	10 to 12 inches (25 to 30 cm) long
RANGE:	Thailand, Southeast Asia, Australia (Indo-Pacific and Oceanic waters)
HABITAT:	Mangrove swamps and river mouths
STATUS:	Least concern

Banded archerfish thrive in coastal waters, swimming between fresh water and salt water. Yet often their insect prey live on land. When a banded archerfish spots a crunchy grasshopper on a leaf above the water, how does it snag its prey? By spitting, of course! Like aiming a Super Soaker, the banded archerfish spits a powerful jet of water at its prey, knocking it off its perch. ***Splashdown!*** A tasty meal. If it misses its target, the banded archerfish can shoot up to seven streams with one mouthful of water. The fish changes the shape of its mouth to aim its spit. The jet of water stings like an insect bite.

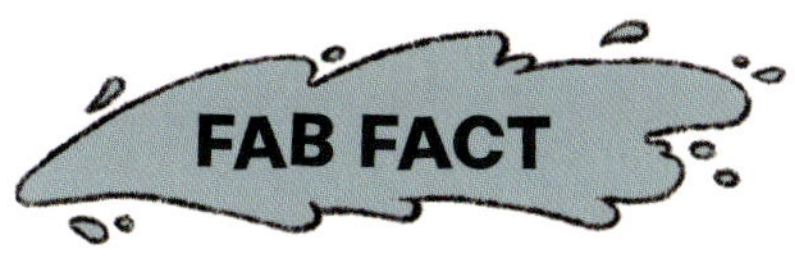

The banded archerfish's genus name, *Toxotes*, is Greek for *archer*, while the species name *jaculatrix* is a Latin word for a female javelin thrower.

Spitting Snakes

NAME: **RED SPITTING COBRA**
SPECIES: *NAJA PALLIDA*

SIZE:	2.3 to 3.9 feet (0.7 to 1.2 meters) long
RANGE:	East Africa
HABITAT:	Dry savanna and semidesert areas
STATUS:	Least concern

Many snakes have fangs to inject toxic venom when they bite. But some Asian and African cobras and their cousins, called *rinkhals*, are a double threat: As well as being able to bite, they have developed a long-range defense system. They spit ***jets of venom*** from specialized fangs into the eyes of a predator or prey from as far as 9.8 feet (3 m) away. A small amount of venom hitting the eye can cause extreme pain or even blindness. And the snakes' accuracy is astounding. They can adjust their spray by tracking a predator or prey's eye movements and making small circular motions with their head to produce a spiral spray. They typically prey on frogs, toads, birds, rodents, and other snakes.

When a red spitting cobra spots prey, it can spit venom forty times in two minutes, and up to eight feet away.

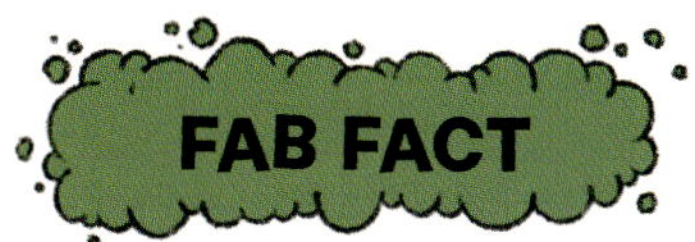

Of 3,500 snake species around the world, only 600 (or less than 25 percent) are venomous.

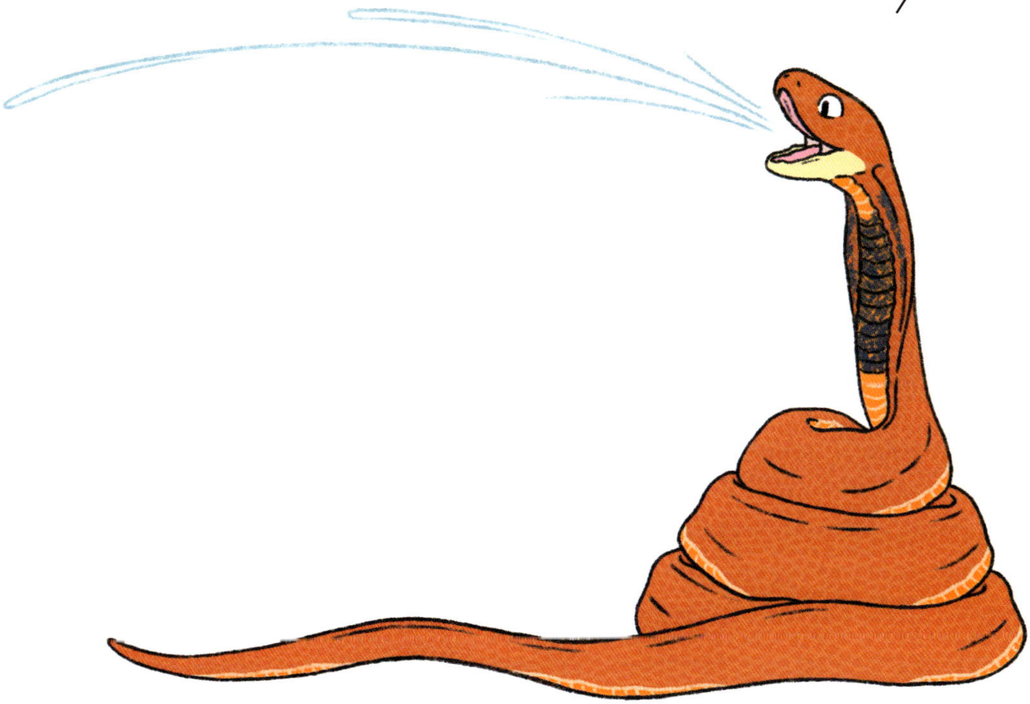

Bull's-eye!

Snake Scientist

Taline Kazandjian, Molecular Evolutionary Biologist

Have you always liked snakes?
I've loved reptiles since I was a child. I don't know where I get it from because my parents don't like snakes. But when I was two or three years old, they took me to an animal show and they had to stop me from crawling over to a big Burmese python.

Have you ever been afraid of snakes?
No. Biting or striking makes me jump a bit, but I'm much less afraid of snakes than I am of public speaking. I don't know why!

Tell us about your educational journey.
I studied zoology in college and earned a Ph.D. in snake venom evolution.

Describe your reptile research.
I work at the Center for Snakebite Research and Interventions at the Liverpool School of Tropical Medicine in the United Kingdom. In one study, we tried to determine how and why some cobras evolved the ability to spit venom as a means of defense. We also study the evolution of snake venom and how snake venom might cause a blood clot or prevent blood from clotting, destroy cells, or affect nerves. My coworkers on the medical team may use our research to develop treatments for snake bites.

Where do you study snakes?
Mostly I work in the lab or at my computer, but sometimes I travel into the field to collect venom from snakes. Recently, I traveled to Namibia, Africa, to study viper venom.

Do you wear protective gear in the field?
We wear goggles, gloves, thick socks, and hiking boots, because if you accidentally tread too close, the most likely place that a snake will bite is near your foot. We also carry hooks to hold the snakes as far away as possible.

Why is studying snakes important?
Snakes are a fascinating piece of the grand puzzle of life and just as important as studying mammals or birds. Some snakes get a bad rap because they bite a lot of people. They can

cause injury and sometimes death. Snake bites are a medically neglected tropical disease. They occur in some of the poorest areas of the world, in Sub-Saharan Africa and large parts of Asia. It's important for scientists to develop more effective and hopefully less expensive treatments and to make them available in remote areas in order to save lives.

What's the "rudest" animal behavior you've seen?

The thorny devil lizard shoots blood from its eyes when threatened!

Do you have any pets?

I have a cat, a leopard gecko, and two snakes—a Centralian carpet python named Jasper, which can grow up to three meters (nearly ten feet) long, and a Sumatran short-tailed python named Fawkes. I keep the snakes in separate terrariums so they can't harm each other. About once a month, I feed them each a thawed rat, which I keep in my freezer.

An estimated 5.4 million people worldwide are bitten by snakes each year, with 1.8 to 2.7 million cases of exposure to venom, according to the World Health Organization.

CHAPTER 4

Gulpers, Gorgers, and Piggy Eaters

"Take little bites! Chew your food!" This advice might sound pointless, but the truth is, chewing properly is the first step in digestion. It prevents choking and allows you to swallow your food easily. Chewing also ensures that when small particles of food arrive in your stomach, gastric juices can break them down so that their nutrients can be absorbed by your body. That's how you get energy to stay healthy and grow.

Nevertheless, some creatures in the animal world gulp their food whole. Check out these "rude" behaviors at the beastly buffet.

FOUL FACTS

- Owls often swallow their prey whole. Then they regurgitate (or cough up) a pellet containing undigested fur, bones, feathers, claws, and teeth.
- A ***pelican*** snags prey by scooping up to three gallons (eleven liters) of water in its lower bill, called a *gular pouch*. If it catches a fish, the pelican tips its head forward to drain the water, then throws back its head to gulp the fish whole.

Face Stuffers

NAME: **BURMESE PYTHON**
SPECIES: *PYTHON BIVITTATUS*

SIZE:	Adults are between 10 and 16 feet (3 and 5 m) long, and weigh as much as 200 pounds (90.7 kg)
RANGE:	India and Southeast Asia
HABITAT:	In grasslands, marshes, swamps, rocky foothills, woodlands, river valleys, and jungles with open clearings, and near permanent water sources
STATUS:	Vulnerable

Large snakes, like this python, guzzle entire alligators, deer, pigs, cows, and even spiky porcupines! And forget using a fork. These slithery reptiles have no limbs to pin down their prey. They also lack sharp teeth to rip into food or smooth teeth to chew. Instead, they swallow their prey whole.

Large snakes like pythons are constrictors. They kill their prey by wrapping their strong coils around the critter's body and ***s-q-u-e-e-z-ing*** tightly until the animal stops breathing. Then the snake opens wide. Its jawbones aren't attached to its skull like yours are. They're connected by ligaments, elastic tissue that allows for maximum gape—that nightmarish open mouth.

The snake usually swallows its prey headfirst. It pushes its own windpipe out of its mouth so it can continue breathing while swallowing. Once it gulps the prey, digestive juices get to work, breaking down the meal into nutrients. Some snakes won't need to feed again for several months or a year.

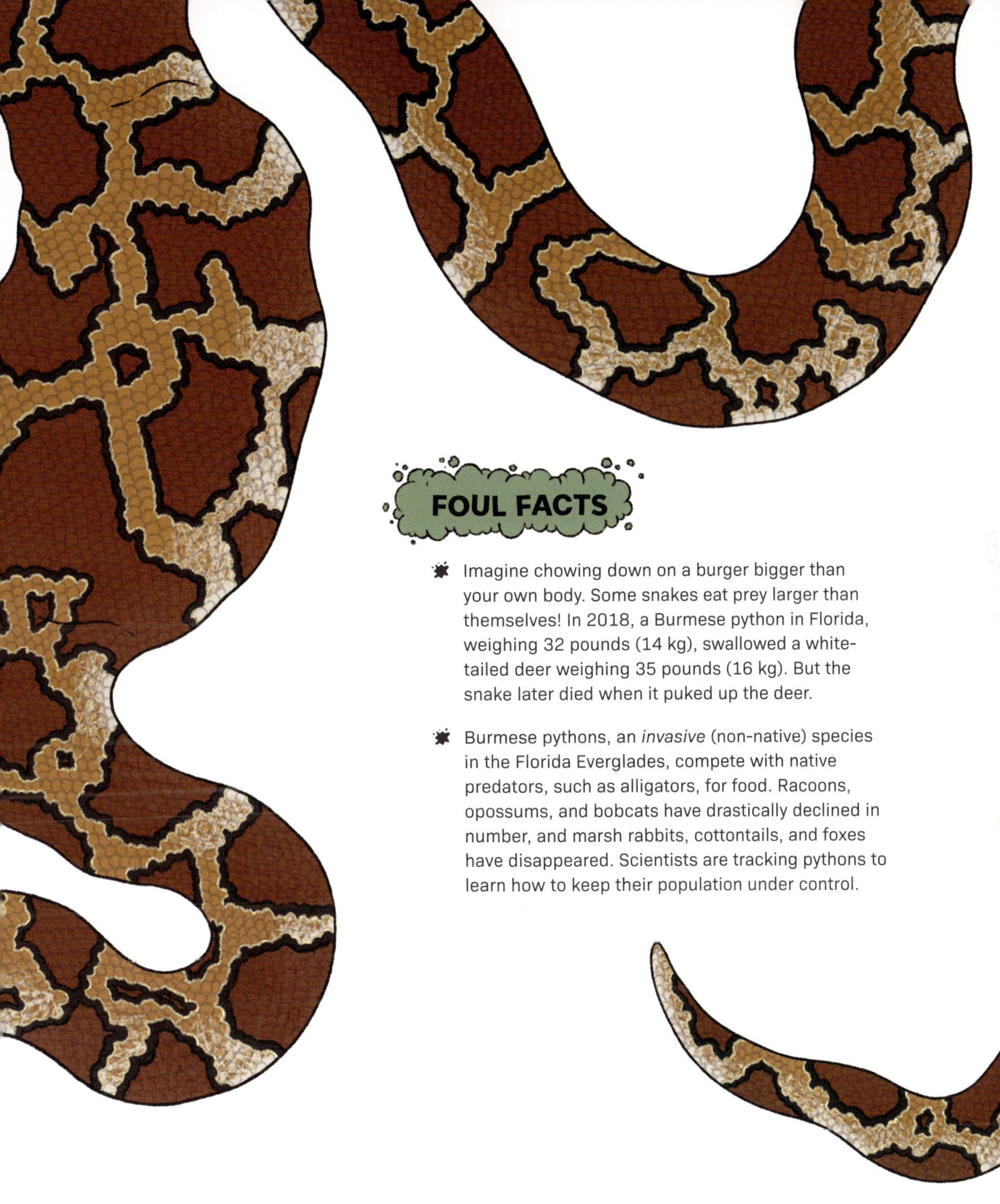

FOUL FACTS

- Imagine chowing down on a burger bigger than your own body. Some snakes eat prey larger than themselves! In 2018, a Burmese python in Florida, weighing 32 pounds (14 kg), swallowed a white-tailed deer weighing 35 pounds (16 kg). But the snake later died when it puked up the deer.
- Burmese pythons, an *invasive* (non-native) species in the Florida Everglades, compete with native predators, such as alligators, for food. Racoons, opossums, and bobcats have drastically declined in number, and marsh rabbits, cottontails, and foxes have disappeared. Scientists are tracking pythons to learn how to keep their population under control.

Gaping Guzzler

NAME: **GULPER OR PELICAN EEL**
SPECIES: *EURYPHARYNX PELECANOIDES*

SIZE:	3 to 6 feet (1 to 2 m) long
RANGE:	Deep seas around the world
HABITAT:	Deep ocean from 500 to 6,000 feet (152 to 1,829 meters)
STATUS:	Least concern

Forget little bites. In the deep, dark sea, you'll swallow any food you can find no matter the size. The gulper eel hunts by opening its massive mouth wide and letting it dangle like a pelican's pouch (that's how the creature got its other name, "pelican eel"). To attract prey, this oddly shaped fish curls its tail over its mouth. At its tip is a *photophore,* a glowing or *bioluminescent* organ that shines a pink light. Fish or shrimp are lured by the glow. ***GULP!*** The eel snaps its jaws shut and swallows its prey whole. The creature's big mouth also scoops up anything in its path, gorging like a hungry bulldozer.

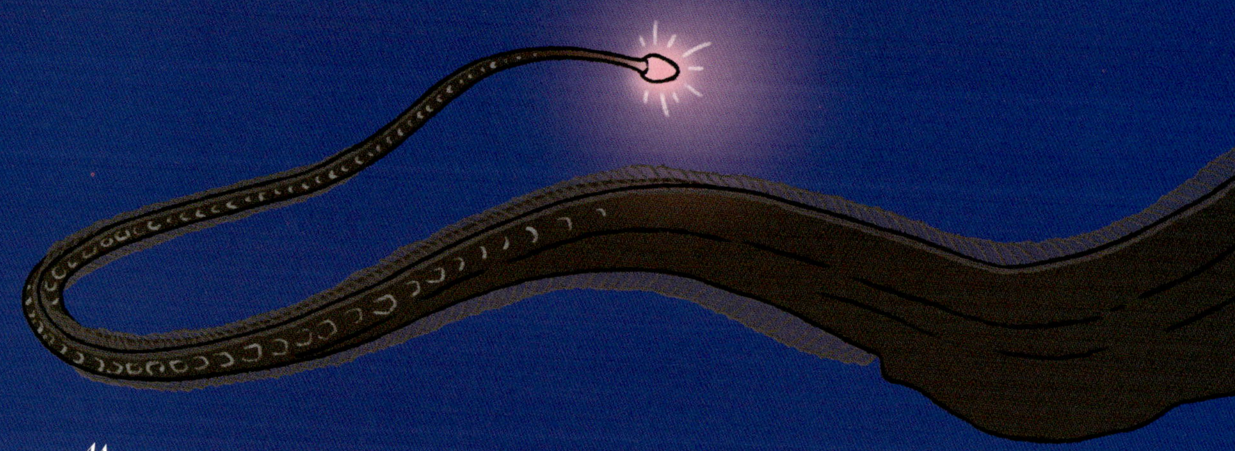

FOUL FACT

In 2018, the crew of the exploration vessel *Nautilus* caught a rare sighting of a gulper eel in the deep sea. Its closed mouth ballooned like an inflating underwater airbag—possibly to scare off predators.

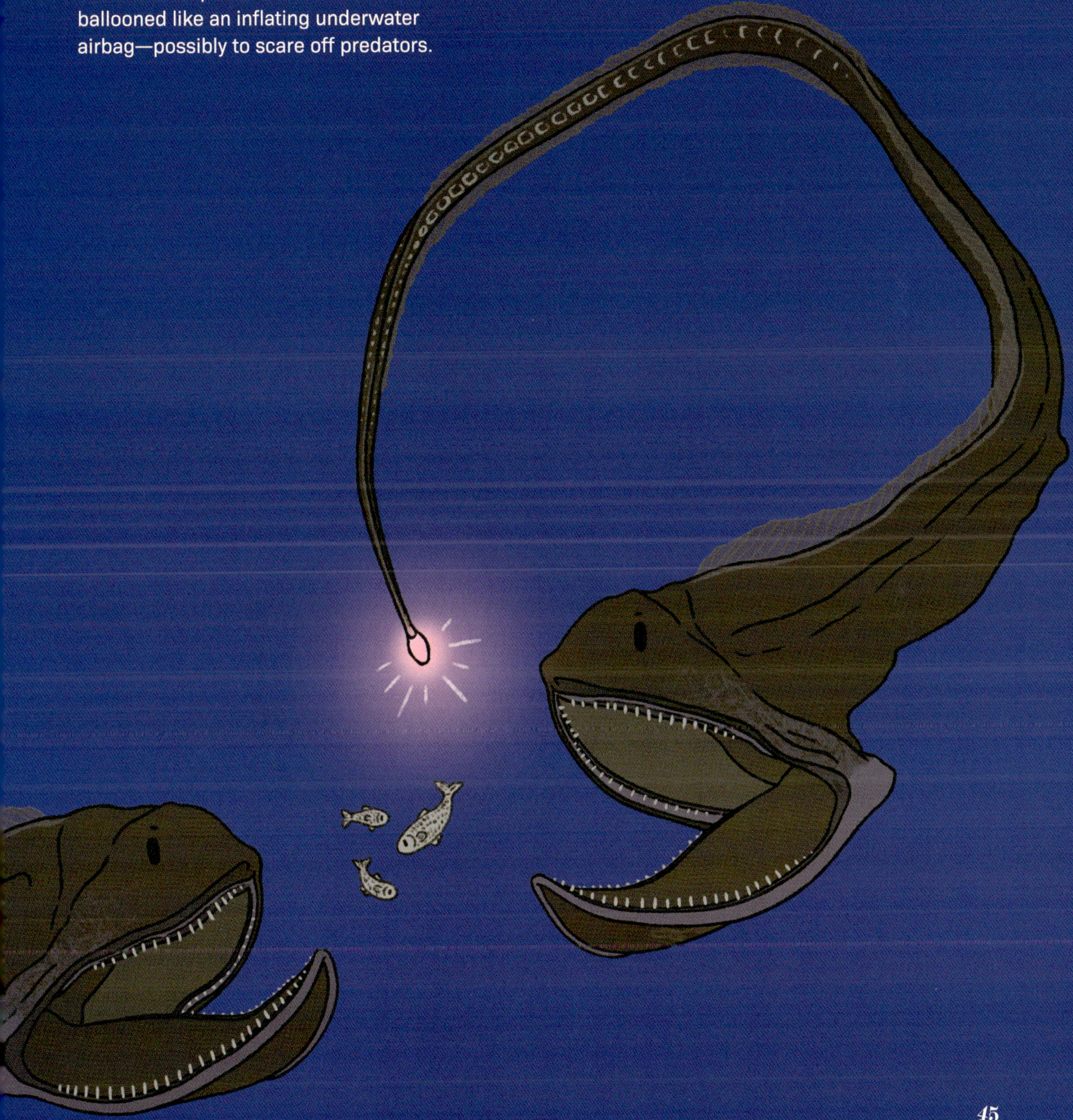

Drooler

NAME: **KOMODO DRAGON**
SPECIES: *VARANUS KOMODOENSIS*

SIZE:	Females up to 6.6 feet (2 m) long, males up to 9 feet (2.7 m) long; adults weigh 150 to 300 pounds (68 to 136 kg)
RANGE:	Indonesia's Lesser Sunda Islands, including Komodo Island
HABITAT:	Tropical forests and savannas
STATUS:	Endangered

Komodo dragons have some of the world's worst table manners. For starters, this giant lizard sticks out its long, forked, yellow tongue to "smell" and track prey. The tongue taps a special organ at the roof of its mouth (called a Jacobson's organ), which identifies what's for supper. A Komodo dragon overpowers its prey by using its claws, teeth, venom, and strong tail to knock them down. When the predator catches its prey—say, a deer, pig, or water buffalo—the lizard clamps down on its throat and tears it to shreds with sixty jagged teeth, each one inch (2.5 cm) long.

Strong muscles in the dragon's throat and jaws allow it to quickly devour big chunks of meat. An expanding stomach means it can snarf 80 percent of its body weight in a single meal. That's why some Komodos eat only about once a month. Sometimes, when it's done—***GAG!***—it upchucks a pellet of the animal's hair, horns, and teeth covered in smelly mucus. You are now excused!

FOUL FACTS

- Glands in the mouth of a Komodo dragon produce ***toxic venom*** that prevents blood from clotting, which can cause its victim to bleed more quickly.
- About three thousand to four thousand Komodo dragons live on five islands in Indonesia, where they are threatened by illegal collection, loss of prey and habitat, and rising sea levels due to climate change. Komodo National Park in Indonesia was established in 1980 to protect this endangered species from poachers.

YUCKY OR USEFUL?

Komodo dragon ***blood*** might be a source for new medications. Researchers have developed a compound called DRGN-1, inspired by a chemical in Komodo dragon blood. DRGN-1 kills drug-resistant bacteria and some fungi. It may also promote wound healing.

Komodo Dragon Expert

Matthew Evans, Herpetologist

How did you become interested in *herpetology*, the study of reptiles and amphibians?
I was always interested in animals as a kid. I grew up near woods and streams. I collected snakes and amphibians and brought them home, and my mom would make me release them.

Describe your education and work.
In college, I studied ecology and herpetology. I knew that wearing a lab coat and looking into a microscope wasn't for me. Instead, I was interested in field biology, as well as conservation and education. Now I work as the assistant curator of herpetology at the Smithsonian's National Zoo and Conservation Biology Institute in Washington, D.C. I work with some of the most critically endangered species on the planet, including Komodo dragons.

How cool are Komodo dragons?
Komodo dragons are iconic! They are the world's largest living lizards. We have three at the National Zoo: Murphy, a twenty-six-year-old adult male; Marigold, a five-year-old female; and Onyx, a three-year old male. All were hatched in zoological institutions.

How do you take care of Komodos?
We participate with other conservation organizations and zoos in a Komodo Dragon Species Survival Plan. Our Komodos live in separate enclosures to minimize aggression among them. We feed these *carnivores* (meat eaters) with five-foot-long tongs so we don't get between them and their food. (That could be dangerous!) Their diet includes defrosted chicken, rats, mice, and rabbits. Komodo dragons rip off large chunks of meat or swallow their food whole. Sometimes we feed them the leg of a cow. They may gorge about twenty pounds of meat in a day.

Are you ever afraid of them?
We train them so they become accustomed to us. The safety of the staff as well as the Komodo dragons is paramount. The training

is hugely important for getting the dragons to shift into crates and holding areas on demand. We establish a checkup and feeding routine. We have a two-person safety rule for entering their enclosure. We wear boots and long pants, never shorts (Komodos have a strong sense of smell). When they're stressed, Komodo dragons clearly show it through their body posture and breathing. These are usually good signs to back off and give them space.

How do Komodos behave in the wild?

They are eating machines. They tend to eat small mammals and other reptiles, but they are also built to take down larger animals, like deer and water buffalo. Komodos have unique adaptations that make them voracious predators. For example, they have sixty teeth that are sharp on one side and have a serrated edge on the other, similar to a shark's teeth. Their thick muscular necks allow them to move their heads side to side, aggressively using those specialized teeth to rip into flesh and muscle. Glands in their lower jaw produce venom that prevents blood from clotting, so their prey go into shock from blood loss.

What are some "rude" Komodo behaviors?

When Komodo dragons are excited about food, they drool a lot. When they kill a large animal, like a water buffalo, many Komodos are attracted to the smell. The largest Komodo dragons are the most dominant bullies and not keen on sharing. Smaller dragons will move away if larger dragons show up.

CHAPTER 5

Bullies, Hoggers, and Thieves

Since you were a little kid, you were taught to follow rules: Use your manners, share, wait your turn, be kind, keep your hands to yourself, and treat others with respect. What could be more rude than snatching someone else's lunch? Or worse, imagine moving into someone else's home without their permission! Without rules or boundaries, human society would go ***kablooey!*** In the animal world, however, many creatures are built to bully, hog, and steal in order to survive.

A female ***cowbird*** lays her egg in another bird's nest, hiding it so that the unsuspecting female bird will care for the egg and raise the cowbird's chick as her own. Once the cowbird chick hatches, though, it becomes a natural-born bully, pushing other eggs out of the nest and growing to be twice the size of the other chicks.

Look out below!
!

Sea Pirates

NAME: **ARCTIC SKUA**
SPECIES: *STERCORARIUS PARASITICUS*

SIZE:	Bodies are 16 to 18 inches (41 to 46 cm) long, with wingspans of 43 to 49 inches (109 to 124 cm)
RANGE:	Breeds in northern North America, Europe, and Asia; spends winters in tropical and southern ocean waters
HABITAT:	Tundra and coastal marshes in the Arctic; open ocean in winter
STATUS:	Least concern globally; endangered in Iceland

Arctic skuas are seabirds that have earned a reputation as pirates of the sea. They steal most of their meals from terns, puffins, and other seabirds. Skuas tend to attack in midair, forcing victims to drop their catch. Sometimes these ***feathered bullies*** will gang up on their victims. They're also double crossers, stealing prey from each other!

A related species, South Polar skuas, swipe penguin eggs and chicks. They have to watch out for adult Adélie penguins, though. Those penguins are much bigger than the skuas (skuas weigh about three pounds, to the penguin's ten to twelve pounds). If a penguin takes down a skua, pirate might as well walk the plank.

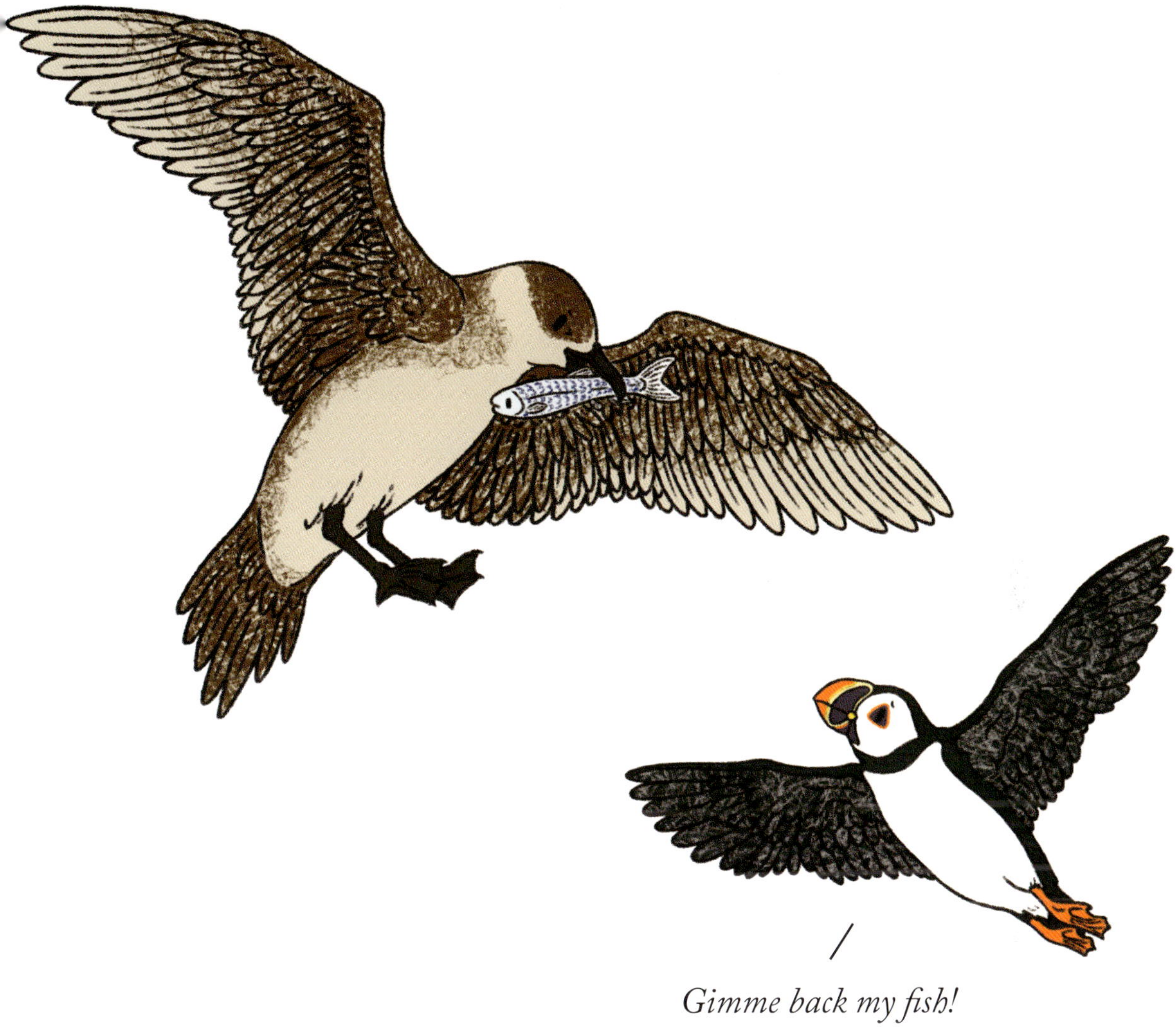
Gimme back my fish!

FOUL FACTS

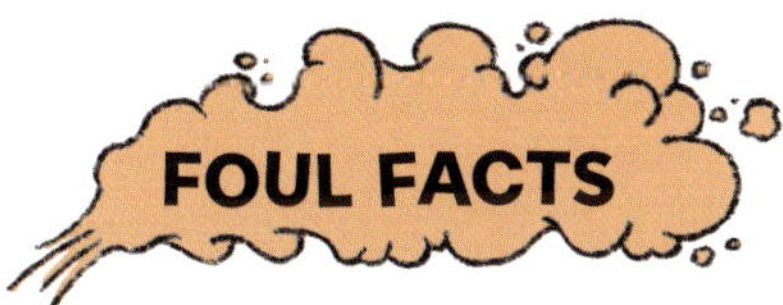

- In North America, these squawking swashbucklers are known as *parasitic jaegers*. *Parasitic* means they live off of other creatures. *Jaeger* means *hunter* in German.
- *Kleptoparasitism* is the term for stealing food from other species.

Bold Bandit

NAME: **HONEY BADGER**
SPECIES: *MELLIVORA CAPENSIS*

SIZE:	Adults are 29 to 38 inches (73 to 96 cm) long and weigh 13 to 30 pounds
RANGE:	Sub-Saharan Africa, Arabian peninsula, India, Nepal, Pakistan
HABITAT:	Wide variety, including tropical and subtropical forests, open woodlands, grasslands, arid steppes, rocky hills, and deserts
STATUS:	Least concern, though its overall population is decreasing

Don't be deceived by this critter's sweet name. The honey badger was listed by the *Guinness Book of World Records* as the world's most fearless creature. A member of the weasel family, it boldly ***picks fights*** with porcupines, hyenas, lions, leopards, and pythons. Despite being named for its preference for honey and honeybee larvae, one-quarter of its diet consists of venomous snakes! Armed with thick skin and sharp teeth and claws, it fights off predators much larger than itself. To rest, it rapidly digs a den in the ground, a tree trunk, or termite mound. If feeling lazy, though, the badger will make itself at home in the dens of other animals, like aardvarks, foxes, mongooses, and springhares.

- When annoyed, a honey badger lets off a skunk-like stink bomb from glands near its rear end. Its message: *Leave me alone!*
- Despite their name, honey badgers are not related to badgers. Instead, they're related to one of the world's most ferocious mammals, the wolverine.
- Honey badgers are also known as "ratels," because they let off a rattling sound when backed into a corner and ready to fight.
- A honey badger's skin is thick and loose, like a baggy jacket. It's even thicker than a buffalo's skin, and a buffalo is fifty times its size. The loose skin allows the honey badger to twist and turn when attacked by a predator, enabling it to swipe the predator with its oversized claws.

Stone Swipers

NAME: **CHINSTRAP PENGUIN**
SPECIES: *PYGOSCELIS ANTARCTICUS*

SIZE:	Adults average about 28 to 30 inches (71 to 76 cm) in length and weigh about 7 to 13 pounds (3 to 6 kg)
RANGE:	Western Antarctic Peninsula, South Atlantic islands
HABITAT:	Onshore colonies, icebergs, and pack ice
STATUS:	Least concern

With their pink feet, little black caps, and bands of black feathers under their chins, these chinstrap penguins may look adorable. But beware! When these cuties gather in massive colonies and build nests, their behavior can get nasty. Males sometimes steal stones from females to build their own circular nests. But that's not how they got the nickname "***stonebreaker penguins.***" When they get annoyed, they let out a screech so loud it's like breaking stones. One saving grace: Males and females share egg-sitting duties. At least they exhibit *some* good manners.

FOUL FACTS

- The largest colony of chinstrap penguins lives on the uninhabited South Sandwich Island of Zavodovski in the South Atlantic Ocean. At the base of an active volcano live six hundred thousand breeding pairs of penguins. No wonder they fight over stones!

- Chinstrap penguin populations have plummeted by half recently on the western Antarctic Peninsula, possibly due to climate change. Melting sea ice disrupts the growth of krill, the chinstrap penguins' source of food. In some colonies, chinstraps are being replaced by gentoo penguins, which have a more varied diet.

- Nesting chinstrap penguins nod off in brief "microsleeps" that average only about four seconds each. However, since they nap about ten thousand times a day, they get eleven hours of slumber. Scientists think this unusual sleep pattern is an adaptation to living in a big, noisy group.

Fowl Follower

Noah Strycker, Ornithologist (Bird Scientist)

Were you interested in wildlife as a child?

I grew up on twenty acres of forest outside of Eugene, Oregon. Bears, bobcats, mountain lions, and all kinds of other critters live in those woods. So I've always had a close connection to nature.

How did you decide to study birds and animal behavior?

When I was in fifth grade, my teacher put a bird feeder on our classroom window. Each time a new bird arrived, we'd try to identify it. I thought it was so cool—and still do!

What is your educational background?

I studied field biology in college, and later researched chinstrap penguins in Antarctica for my master's degree. One of the best decisions I ever made was to take a year off between high school and college to gain field experience. That year, I searched for bird nests in Panama, volunteered at a wildlife refuge in Oregon, and traveled to East Asia to help design an ecotourism program that centers on an awareness of the environment and the local community.

What does your day-to-day research look like?

In Antarctica, we monitor penguin populations by counting their nests. It helps to break the colony into smaller landmarks. I use a hand clicker and field notebook to keep track of numbers. For very large colonies, our research team also uses drones and even artificial intelligence to interpret aerial photos. But usually, it's just me standing on a rock, counting penguin nests one by one. Over time, this provides valuable information about how well the penguins are doing.

Do you also conduct research in a lab?

All of my research has been field-based rather than in a lab. A lot of ornithologists now use cutting-edge techniques to study genetics, but field-based research is still important. Nothing will ever replace someone with binoculars looking at wild birds.

Which birds fascinate you most and why?

I'm especially interested in behaviors that birds share with humans. Birds really do think and feel, just like we do. I'm fascinated by how birds experience thoughts and emotions.

traveled to forty-one countries on all seven continents! (A Big Year is a personal challenge among birders who attempt to identify as many species of birds as possible by sight or sound within a single calendar year and within a specific geographic area.)

Do you have any pet birds?

I've always been a cat person. My family has a tabby cat named Bernstein. He likes to watch birds from our kitchen window.

What's the "rudest" animal behavior you've witnessed (other than by humans)?

In penguin colonies, the nests are packed close together. And when a penguin poops, it deliberately aims away from its own nest. I've watched penguins shoot their neighbor with poop, straight in the face! The poor pooped-on penguins have to sit on their nests until their mates return. Then they can go for a swim and get clean.

You broke a world record! Tell us about it!

I set the world record for most bird species ever seen in one year: 6,042 species during my 2015 round-the-world Big Year, when I

CHAPTER 6

Public Poopers

Splat! A bird lets loose on your head. ***Plop!*** A dog leaves a load on the sidewalk. ***How rude!*** What's the deal with doo?

Every animal poops! It's one of the body's ways of dumping waste. In humans, undigested food travels from the stomach, through the small intestine (where some nutrients are absorbed) to the large intestine, or colon. That's where stool (aka feces or poop) forms. Poop contains water, undigested food, dead cells, and bacteria. Different foods and bacteria in your gut can cause changes to the texture, smell, and color of poop. Passing blue poops? Chances are you drank a blue drink.

Every animal (other than, say, a potty-trained human) poops in public. But did you know that some animals put their droppings to good use? Check out what these animals do with doo.

FOUL FACT

Koala babies, called *joeys*, nuzzle their mom's rear end and eat her poop pellets, as well as a runny, protein-rich substance called *pap*, which is full of microbes that help the babies digest eucalyptus leaves, the food they'll eat as adults.

Dung Dinner

NAME: **GUINEA PIG**
SPECIES: *CAVIA PORCELLUS*

SIZE:	8 to 16 inches (20 to 41 cm) in length, and 1 to 3 pounds (454 to 1,361 g) when fully grown
RANGE:	Domesticated (originated from the Andes of South America)
HABITAT:	In people's homes; their wild cousins live in forests, scrub deserts, and savannas in South America
STATUS:	Least concern

CAUTION: This *might* be the grossest behavior yet. Some animals—like mice, guinea pigs, chinchillas, capybaras, rabbits, and hares—feast on their own feces (or dine on their own dung, or gorge on their own guano . . .). It's nature's way of providing their bodies with nutrients that might have passed through them the first time around.

After eating lots of plants, many of these animals poop two kinds of pellets—dry, hard turds that have little nutritional value, and soft pellets called *cecotropes* that are packed with nutrients like nitrogen, protein, and vitamins. For these critters, consuming their own poop isn't just normal; it's necessary!

The scientific term for dining on dung is *coprophagy*.

Poo-ble Beach

NAME: **PARROTFISH**
FAMILY: *SCARIDAE* (MORE THAN 90 SPECIES)

SIZE:	Adults range from less than 1 foot to 4 feet (0.3 to 1.2 m) long
RANGE:	Subtropical and tropical oceans worldwide
HABITAT:	Coral reefs
STATUS:	Least concern

On a dreamy tropical beach, you're about to build a sandcastle when you discover this: All that beautiful white sand is actually parrotfish poop! In warm tropical seas, these colorful fish use their sharp, beak-like teeth to gnaw off chunks of coral covered with algae. They grind up the rock-hard coral with a thousand teeth lined up in fifteen rows.

Parrotfish don't have a stomach; instead, their meals pass straight through their small intestine. Coral is made up of calcium carbonate, which the fish can't digest. ***POOF!*** An explosion of ground-up white sand shoots out of the parrotfish's rear end. A large parrotfish can poop out 2,000 pounds (907 kg) of sand per year. That's the weight of a grand piano or one really enormous sandcastle.

Hungry for Poo

NAME: **HIPPOPOTAMUS**
SPECIES: *HIPPOPOTAMUS AMPHIBIUS*

SIZE:	On average, 10.8 to 16.5 feet (3.3 to 5 m) long and up to 5.2 feet (1.6 m) tall; males weigh 3,500 to 9,920 pounds (1,588 to 4,500 kg) and females weigh 3,000 pounds (1,361 kg)
RANGE:	Africa
HABITAT:	Rivers, lakes, and swamps
STATUS:	Vulnerable

Hippos hang out in rivers and lakes for sixteen hours a day. At night, they lumber onto land and feed on gobs of grass—about eighty to a hundred pounds in one meal. (That's like snarfing as many as four hundred burgers in one sitting.) Then the hippos head back to the water—to lounge, digest, and poop.

Researchers have discovered that, under normal conditions, hippo dung actually plays an important role in the web of life. It feeds the fish in this *ecosystem,* a community of interacting organisms. Scientists traced a chemical trail from the hippos' behinds to the tissue of fish in a river in Kenya. By eating grass and then pooping in the water, the hippos are providing a fish-food-delivery service, transporting nutrients from land to the fish in the water in their poop. In nature, nothing is wasted—not even poop!

However, when water levels are low during the dry season, or when rivers are drained for farming, nutrients in hippo poop could cause algae to bloom and lead to fish die-offs. Fewer fish can mean less food for other animals and people, too. It's a delicate, poopy balance.

- Along Kenya's Mara River, about four thousand hippos gather in pools. They collectively dump about 9.3 tons (8,500 kg) of poop each day. When water levels are low and the concentration of poop is high, microbes in the poop take up oxygen and release chemicals that are lethal to fish. Rainstorms "flush" the poop downstream, killing off more fish.
- Hippos mark their territory using *dung middens*, piles of waste where they repeatedly poop. They sniff middens to determine whether there are friends or enemies in the area.
- Hippos' large canine teeth are made of ivory, the same material as elephant tusks. This puts them at risk from poachers, who illegally kill them for their canines.
- A group of hippos is called a *bloat*.

Poop Scientist

Scott "Scat Man" Burnett, Wildlife Ecologist

How did you step into the study of scat (animal poop)?

I grew up in Brisbane, Australia, and my backyard bordered on bushland reserves, where there is lots of wildlife. I loved seeing bearded dragons (a type of lizard) and beautiful snakes. Later, I earned a Ph.D. in wildlife ecology, the study of the relationship between living things and their ecosystems. Studying scat may seem gross, but it's part of a tool kit for scientists to learn about animals and their environment—what they eat, their health, which animals share their territory, and more.

What clues do you find in poo?

Poop is like golden treasure for scientists because it gives us clues to *conservation*, protecting species in their environment. For example, I studied an endangered species in Australia called a spotted-tail quoll, which is related to opossums. These small mammals are hard to find in nature, but by studying their scat, we can discover what they eat, if something is missing in their diet, where they fit in the *food web* (interconnecting food chains that show relationships among animals), and if there are ways we can protect their environment.

How do different species scatter their scat?

Some animals poop on the go. But sometimes animals poop in latrines! These are like public toilets for animals (without the toilets). When studying quolls, I discovered a ten-kilometer (six-mile) stretch of road they were using as a latrine. By studying diverse doo, you can find out which other animals use the same latrine. Animals use latrines for social communication—like Fecesbook (instead of Facebook) for animals! They communicate messages through their poop. For example, a pooping female rhino may be marking her territory or letting males know she is ready to mate.

Do you have favorite feces?

Hard to choose, but wombat poops are shaped like cubes! With other scientists, I've explored how wombats make six-sided scat. Patricia Yang, a mechanical engineer who studies biological fluids, investigated the intestines of wombats. The intestines of most mammals (like humans) squeeze out sausage-shaped poops, like squeezing out globs of toothpaste from a tube. But wombat intestines have bands of thicker and thinner muscles. The muscles contract and expand like a rubber band, forming a cubic shape. No one is sure yet *why* wombats produce stackable poop.

Have you studied scat in the sea?

Yes! Blue whales off the coast of Sri Lanka produce fascinating feces. Their pinkish-red poops are like floating gold! By researching the DNA (genetic material) in their poop, marine biologist Asha De Vos has discovered how whale poop helps fertilize the ocean.

The whales feast on krill, small invertebrates that resemble tiny shrimp and swim in huge swarms. When whales poop, they deposit iron in the ocean. Phytoplankton, microalgae that are the base of the aquatic food web, take up the iron. In turn, the krill feed on phytoplankton, and the cycle repeats.

What's the "bottom line" when studying scat?

The truth can be right under your nose! It's a matter of looking at the world a bit differently and using your imagination.

- "Scat Man" Scott Burnett stars in a NOVA/Terra Mater Studios documentary called *Secrets in the Scat* (2022), which follows scientists around the world tracking animal poop.
- In his documentary, Burnett caught up with a cassowary, a giant flightless Australian bird whose poops contain seeds from 238 different plants. On its journey through the rainforest, the cassowary poops and spreads seeds from these plants, keeping the forest diverse and healthy.

CHAPTER 7

Barfers

What could be worse than pooping in public? How about puking your guts out—*literally*? Vomiting is our body's way of getting rid of harmful substances, like food poisoned with bacteria. We also hurl when we have stomach infections caused by bacteria or viruses. These illnesses usually pass in a few days. But the animals shown here upchuck to survive. And some even spew their own insides. If you can stomach this "rude" behavior, read on.

FOUL FACT

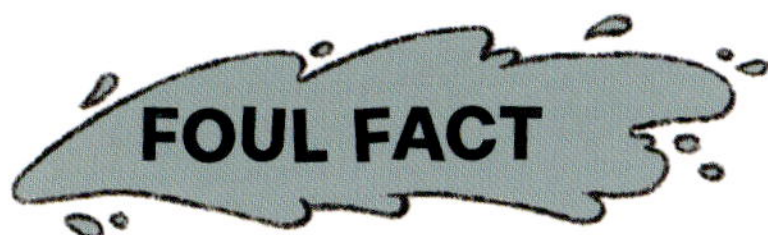

Gastric brooding frogs, which are sadly extinct, had a surprising way of giving birth to frog babies. The mother would swallow her fertilized eggs and then wouldn't eat for six weeks while her babies grew inside her. Finally, she gave birth by puking out her little froglets in a projectile burst. No other animal has been known to do this.

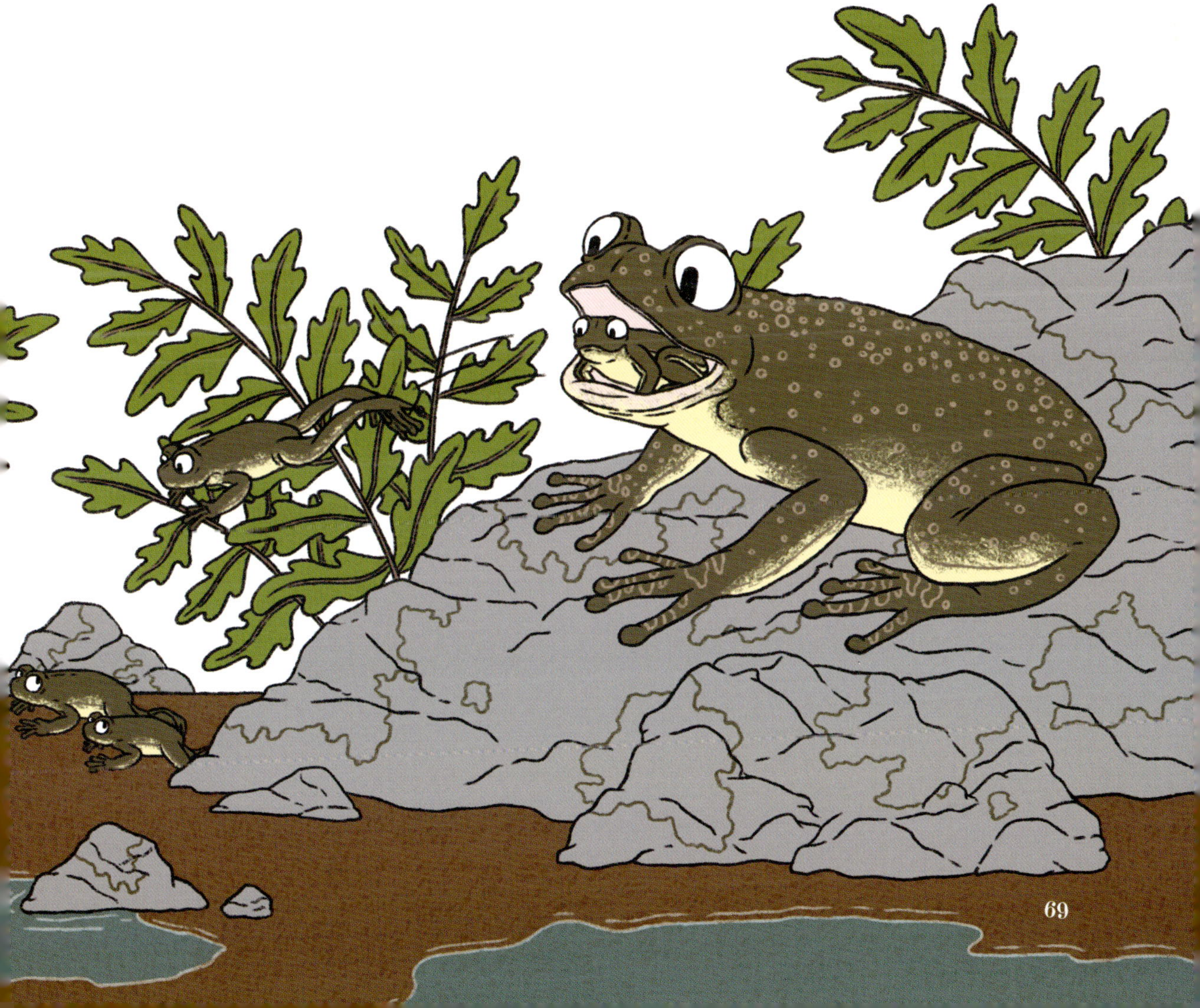

Stomach Turner

NAME: **SEA CUCUMBER**
CLASS: *HOLOTHUROIDEA* (1,250 SPECIES)

SIZE:	0.75 inches to 6.5 feet (0.019 to 0.17 m) long
RANGE:	Tropical, subtropical, and temperate oceans worldwide
HABITAT:	On the seafloor near coral reefs, seagrass beds, and other fixed habitats
STATUS:	Vulnerable or endangered

Sea cucumbers outstrip frogs and sharks when it comes to tossing their cookies. (Note: Sea cucumbers are soft-bodied animals, not veggies!) When threatened by a predator, these seafloor-dwelling echinoderms—relatives of sea stars—not only eject their stomachs but also the long, wiggly tendrils of their intestines. It's the cucumbers' way of scaring off predators or possibly tangling them in their tendrils. And if a tendril breaks off, it will grow back.

YUCKY OR USEFUL?

Sea cucumbers have another cool trick: They can change their body texture at will, going from squishy and jellylike to firm in a matter of minutes. This helps them wedge into tight spaces to hide from predators. Scientists are investigating possible applications in tissue engineering and soft robotics.

FOUL FACT

Discarding an organ to save your skin is known as *autotomy*.

Spewing Bird

NAME: **TURKEY VULTURE**
SPECIES: *CATHARTES AURA*

SIZE:	Bodies are 26 to 32 inches (0.6 to 0.8 m) long, with wingspans of up to 6 feet (1.8 m) and weight of 3.5 to 5 pounds (1.6 to 2.3 kg)
RANGE:	North, Central, and South America
HABITAT:	Subtropical forests, shrublands, pastures, and deserts
STATUS:	Stable

You might see these big birds on the side of the road. Turkey vultures mostly feed on carrion, the dead and rotting bodies of animals. Vultures locate dead animals by smell and sight. If threatened, turkey vultures have a rude way of repelling predators: They vomit their food, which they can spew as far as ten feet (three meters). If another animal harasses them, they might even upchuck directly onto that animal. Vulture babies do the same thing. (Well, if we're being honest, human babies puke on other people too, but usually not on purpose.)

- The turkey vulture's genus name, *Cathartes,* means *cleaner*, because they clean up carcasses.
- On hot days, turkey vultures poop on their feet to cool them off. Pooping on their legs might also kill bacteria, as their digestive juices have antibacterial properties. This is why turkey vultures don't get sick after eating rotting meat.
- Eating roadkill might seem gross, but turkey vultures are winged vacuum cleaners. By eating dead animals, they help prevent the spread of diseases like botulism and anthrax.
- A group of vultures is called a *committee*, *venue*, or *volt*. In flight, they're called a *kettle*, and when eating a carcass, they're called a *wake*.

Because the chicken didn't make it!

Gag Bags

NAME: **JAPANESE TREE FROG**
SPECIES: *HYLA JAPONICA*

SIZE:	1 to 2 inches (2.5 to 5 cm) long
RANGE:	Japan, Korea, China, Mongolia, Eastern Russia
HABITAT:	Forest-like environments, bushlands, meadows, swamps, and river valleys
STATUS:	Least concern

Everyone loses their lunch at some point. It's our body's way of getting rid of toxins like the bacteria in spoiled egg salad. But if a Japanese tree frog feels green (LOL), it doesn't puke. Instead, if it eats something toxic, it throws up its ***entire stomach***! Like blowing a big bubble, the frog turns its insides out. Afterward, it uses its feet to scrape off any spew, then it swallows its stomach whole and hops around as if nothing ever happened.

FOUL FACTS

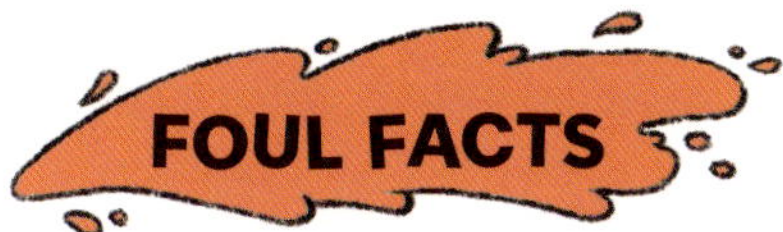

- The scientific term for puking up the stomach is *full gastric eversion*. It's kind of like emptying your pockets by pulling them out of your pants, only much grosser!
- We humans can't eject our stomach. We have an esophageal sphincter, a muscular ring that seals the stomach from the *esophagus*, the tube connecting the throat to the stomach. The sphincter allows you to vomit bad food, but not your entire stomach.

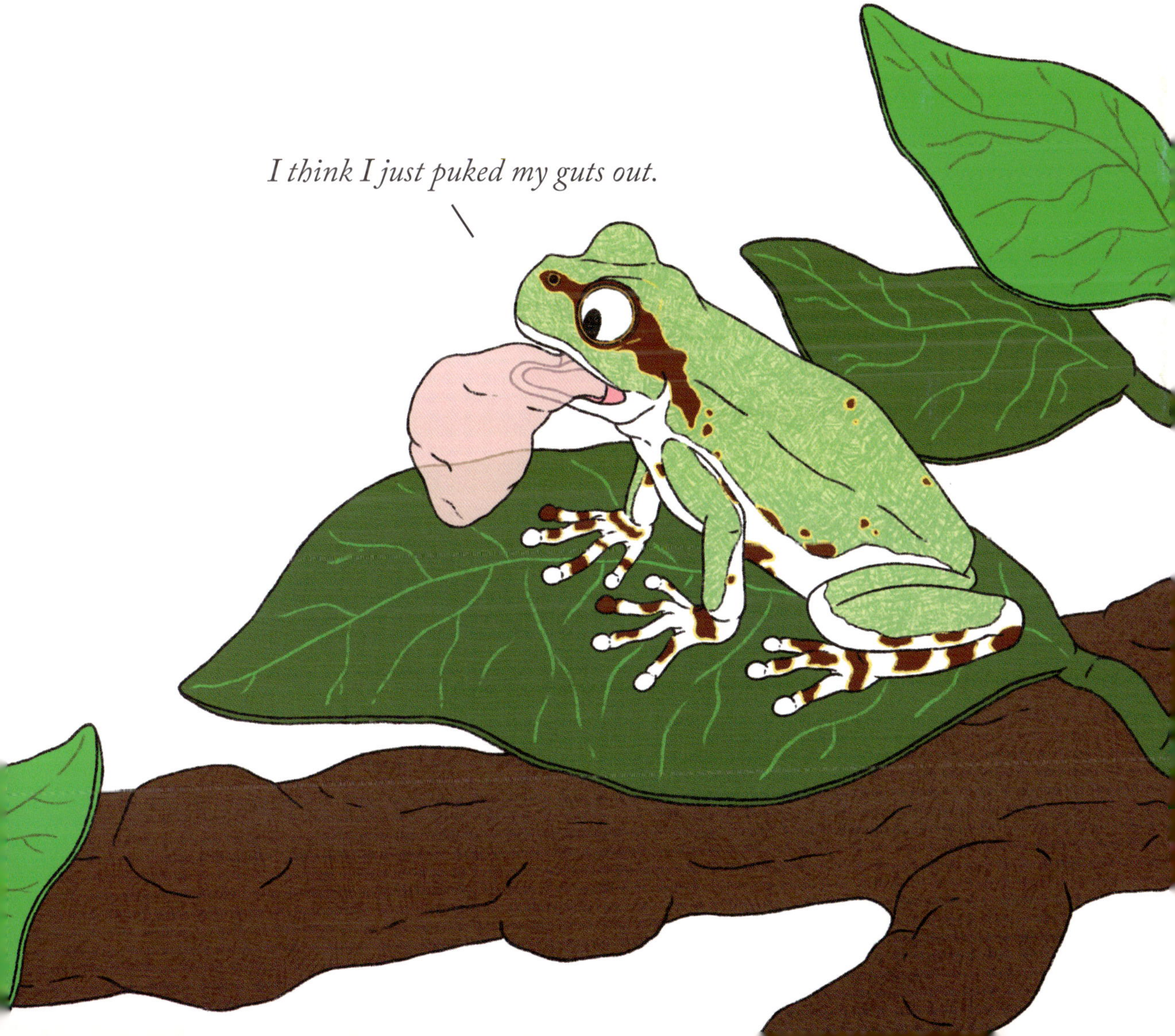
I think I just puked my guts out.

Seasick Swimmer

NAME:	**TIGER SHARK**
SPECIES:	*GALEOCERDO CUVIER*

SIZE:	10 to 14 feet (3 to 4.3 m) long and weight of 850 to 1,400 pounds (386 to 635 kg)
RANGE:	Tropical and subtropical oceans around the world
HABITAT:	Shallow areas around large island chains, including the lagoons and coral atolls found on the coasts of oceanic islands
STATUS:	Near threatened

As apex predators at the top of the food chain, sharks are not picky eaters. They gorge on anything in their path. Sometimes they down some unpleasant side dishes, like bird feathers, turtle shells, lobster claws, and plastic. If a shark is in distress—say, it gets caught in a fishing net—it will hurl its undigested food. Scientists say that by emptying its stomach, the shark will be able to swim faster and make a quicker escape.

Sometimes, like a frog, a shark will upchuck its entire stomach. This helps remove indigestible food, parasites, and mucus. It's like giving their insides a rinse. And great white sharks are known to blow chunks of food just to make room for more. Researchers think this is the shark's way of testing the quality of their meal before going after a better chunk of meat. ***Mmmm! Another bite of whale carcass, please!***

I don't appreciate your shark-asm.
Chews wisely!

Shark Scientist

Yakira Herskowitz, Marine Ecologist

When did you first become interested in sharks?

When I was twelve, I had a counselor at sleepaway camp who loved sharks. At first, I thought that was crazy, but I loved everything else that she taught me, so I kept an open mind. At home, I took a deep dive into reading and learning about sharks. When I was sixteen, my counselor joined me as a chaperone to volunteer at a shark lab in Bimini, Bahamas. I've been hooked ever since.

What is your educational background?

In high school, I started a marine science club that became very popular. In college, I majored in wildlife ecology. In graduate school, I earned a master's degree in marine ecosystems, with a concentration in shark research and conservation.

How do you study sharks?

I study shark behavior and how they interact with other marine species. Sometimes I free dive to observe sharks in the water. (Free diving involves holding your breath underwater. The bubbles from scuba diving can scare away sharks.) My team and I also conduct research by catching sharks and hauling them onto a boat. We quickly take measurements and blood and muscle samples, and then tag them for tracking before releasing them. While on the boat, we keep the sharks alive by inserting a tube into their mouth that pumps water through their gills, which provides them with oxygen.

We also do underwater surveys by lowering a video camera mounted on a pole with bait attached. Later, we spend many hours on a computer observing and recording data on which species swim by. We used this method to study changes in marine ecology in False Bay, South Africa, after great white sharks mysteriously disappeared from that area in 2018.

Have you ever been afraid of sharks?

My fear has been replaced by fascination. Sharks tend to be more afraid of humans. That said, sharks are wild animals, and there are no guarantees about how they'll behave. So it pays to have a healthy awareness of your surroundings when you're in the water.

What "rude" behaviors have you experienced among marine creatures?

Some sharks, like tiger sharks, vomit their entire stomach, a process called *gastric eversion*. I've also seen a shark expel its intestines not through its mouth but from its *cloaca* (rear end). There are a few theories as to why they

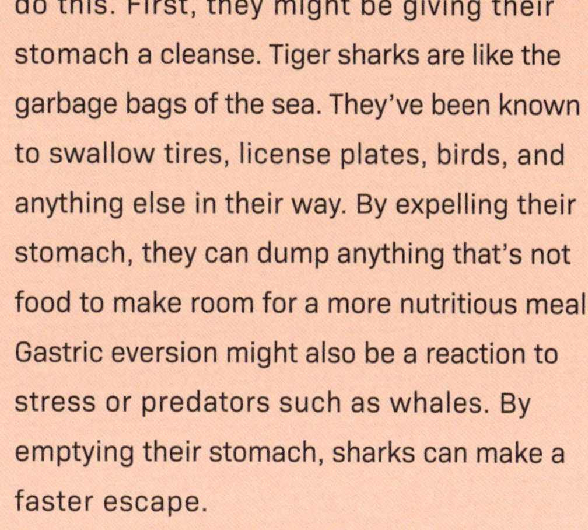

do this. First, they might be giving their stomach a cleanse. Tiger sharks are like the garbage bags of the sea. They've been known to swallow tires, license plates, birds, and anything else in their way. By expelling their stomach, they can dump anything that's not food to make room for a more nutritious meal. Gastric eversion might also be a reaction to stress or predators such as whales. By emptying their stomach, sharks can make a faster escape.

Why is studying sharks important?

Sharks are misunderstood. They are critical for maintaining balance in the ocean ecosystem. Yet seventy to a hundred million sharks are killed each year, often for their fins. (By comparison, only about five or six humans are killed each year as a result of shark attacks.) About 75 percent of shark species are threatened and in need of protection.

- A shark can shed more than thirty thousand teeth in its lifetime. When a tooth falls out, another one quickly grows back.
- An amateur fossil hunter recently discovered 66-million-year-old fossilized shark vomit on a Danish island. Best guess: During the Cretaceous period, a bottom-dwelling shark snarked some sea lilies (a marine animal that looks like a plant), and upchucked the skeletons.

CHAPTER 8

Screechers

Eek! What's that shriek? Sometimes people scream to let off steam or to signal danger, fear, or extreme joy (think roller-coaster ride). There are times when being loud is downright obnoxious. But some people make loud sounds without having control. They're not being rude; their brain is just wired differently.

In nature, animals make loud sounds to be strategic. They roar, howl, croak, and make other ear-piercing sounds to communicate, send out warnings, snare prey, and attract mates. Listen up!

FOUL FACTS

- Normal human conversation is about sixty decibels (dB), the unit that measures the intensity of sound. In October 2000, the record for the loudest human scream was set at 129 dB. But shouting that loud could hurt your ears.
- A ***lion***'s roar can be heard from several miles away. Big cats use their ferocious roars to establish territory and communicate with members of the pride.

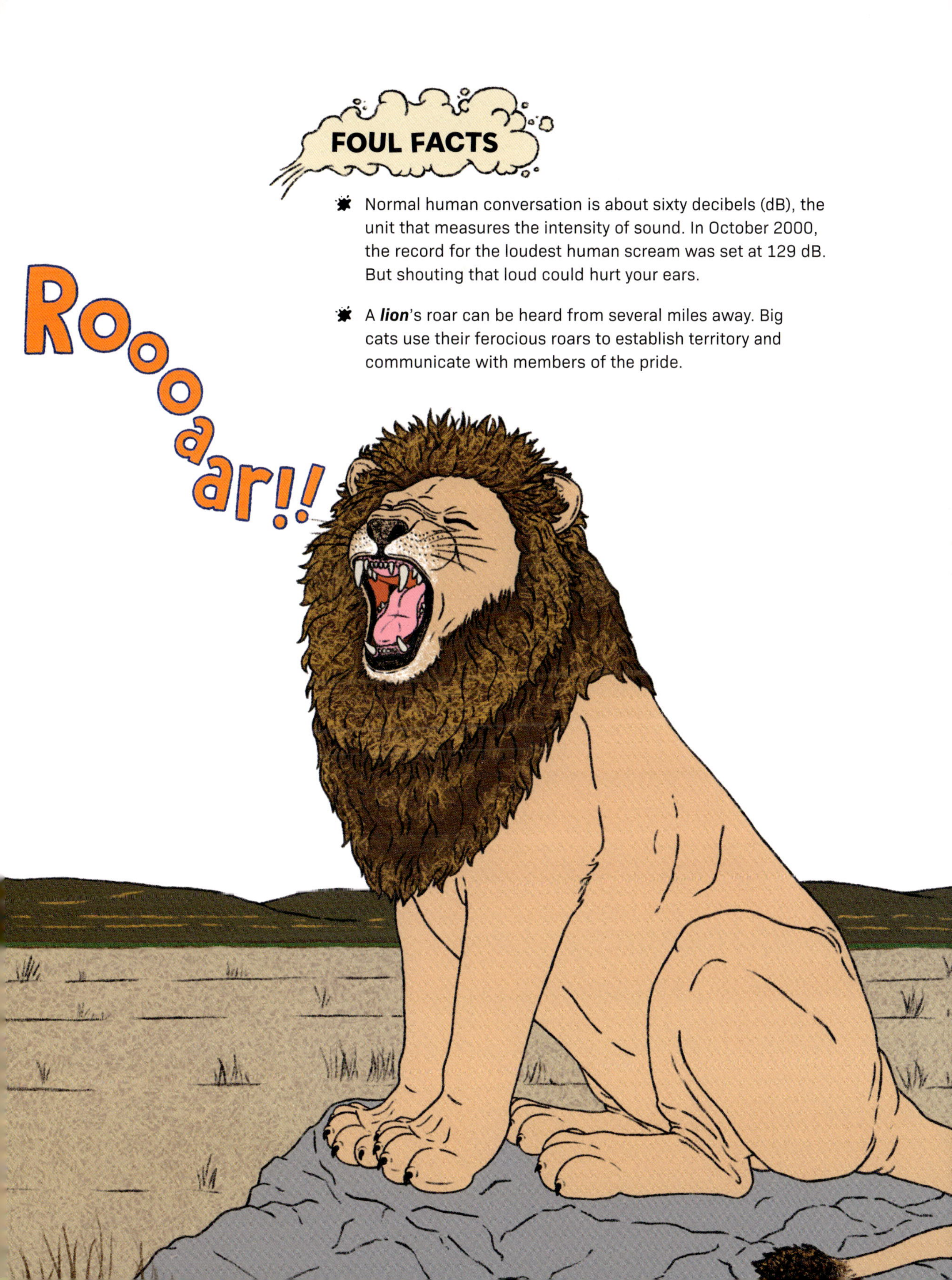

Undersea Jet Engine

NAME: **BLUE WHALE**
SPECIES: *BALAENOPTERA MUSCULUS*

SIZE:	88 feet (27 m) long on average, the largest measured is 110 feet (33 m); weight is 200 tons (181,437 kg)
RANGE:	All oceans
HABITAT:	Open waters
STATUS:	Endangered due to overfishing and climate change

The blue whale, one of the world's largest animals—stretching the length of three school buses—is no quiet giant. At 188 dB, its call is louder than a jet engine's roar (140 dB). A blue whale's pulses, groans, and moans can be heard by other whales a thousand miles (1,600 km) away. A whale might be calling out to find a mate in the vast ocean. However, a blue whale's vocalizations are usually beneath the frequency that humans can hear. Scientists use special devices to record the sounds.

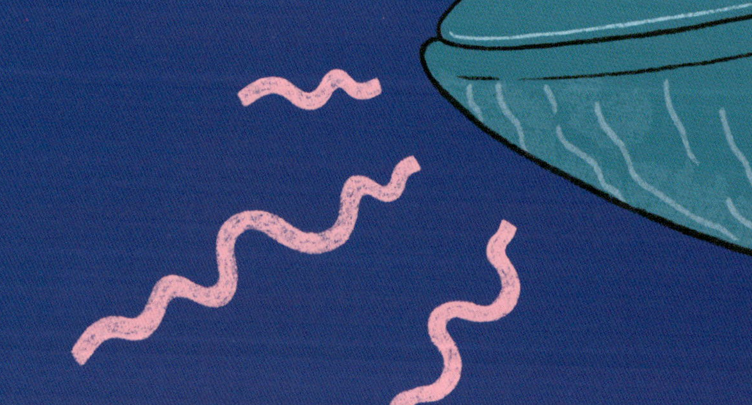

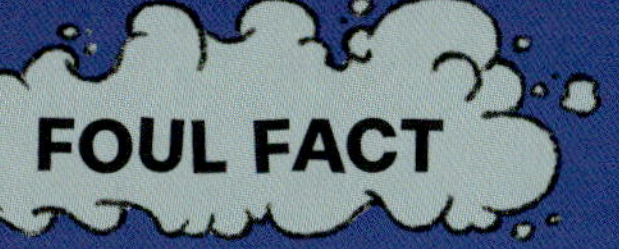

Blue whales aren't just noisy—they're also piggy eaters. A blue whale eats about forty million krill every day. That's about 7,900 pounds (3,600 kg) of food.

Whale, hello there!

Piercing Primate

NAME: **BLACK HOWLER MONKEY**
SPECIES: *ALOUATTA CARAYA*

SIZE:	Head and body are 22 to 36 inches (56 to 91 cm) long, with tail 23 to 36 inches (58 to 91 cm) long. Weight is 12 to 22 pounds (5 to 10 kg)
RANGE:	Central and northern South America
HABITAT:	Tropical rainforests
STATUS:	The black howler monkey in Belize is endangered; other species are threatened

Ear-splitting howls echo through the rainforest. Those scary sounds are coming from howler monkeys hanging out in the trees as far as 3 miles (5 km) away. Males have large throats and specialized vocal chambers that allow them to crank up the volume to 140 dB. They're the loudest terrestrial (land-based) animals in the western hemisphere. At dawn or dusk, their shrieks might be a warning to other monkeys to stay away.

They also use their explosive calls to shout down competitors and attract mates. (Yes, loudness in the monkey world is attractive.) For all their noise, howler monkeys spend 80 percent of their day just chilling in the trees. All that yelling tires them out.

Sonic Snap

NAME: **TIGER PISTOL SHRIMP**
SPECIES: *ALPHEUS BELLULUS*

SIZE:	1.6 to 2 inches (4 to 5 cm) long
RANGE:	Mediterranean and Indo-West Pacific Ocean
HABITAT:	Tropical reefs
STATUS:	Common

One of the loudest creatures on Earth is a tiny shrimp. The tiger pistol shrimp snaps its oversize claw to create a two-hundred-dB sound that's louder than a jackhammer. Luckily, this little noisemaker is under the sea. The human ear can only tolerate sounds as loud as 120 to 130 dB. Sounds any louder could cause pain and possibly deafness.

Here's how the little shrimp makes its big sound: When it spots prey, the shrimp snaps its large claw. The snapping motion shoots out a jet of water at about 60 miles per hour (97 kph). The water moves so fast it creates an air bubble. Within a millisecond, the bubble implodes. It creates a shock wave that makes a blast louder than a whale's call. The sonic snap can stun or kill prey up to 6 feet (2 m) away. The high-energy sound waves also emit bursts of light, which cause the temperature inside the bubble to hit between 7,232° and 8,672°F (4,000° and 4,800°C), nearly as hot as the surface of the sun, or four times as hot as lava. So the sound isn't only loud. It sizzles!

Ear-Splitting Amphibian

NAME: **COMMON COQUÍ FROG**
SPECIES: *ELEUTHERODACTYLUS COQUI*

SIZE:	Adult males, 1 inch (2.5 cm) long; females, 2 inches (5 cm) long
RANGE:	Native to Puerto Rico (now in the Dominican Republic, Virgin Islands, Florida, California, and Hawaii)
HABITAT:	Rainforests and other moist habitats
STATUS:	Least concern

"Co-quí (ko-KEE)! Co-quí!" Meet the coquí frog, the loudest amphibian on the planet. At up to 100 dB, choruses of these tiny male Puerto Rican rainforest frogs create sounds as loud as lawn mowers, motorcycles, or ATVs. The *co* sound deters other males, while the *quí* sound attracts females, who constitute the audience for these singing competitions.

You can catch a concert of these loudmouth frogs at El Yunque National Forest in Puerto Rico, home to eleven out of sixteen species. The frogs are beloved in their home territory, inspiring songs and books and even becoming an unofficial mascot of the island's people and culture.

On average, coquí move only a few hundred meters in their lifetime. But in the 1980s, many Caribbean coquí hitched a ride with nursery plants imported to the Big Island of Hawaii. Because they have no natural predators there, and because they eat all kinds of insects, the coquí population exploded. Now these cacophonous croakers have become a major nuisance. One study found twenty thousand to fifty thousand coquí per acre in Hawaii, more than twice the number in a similar area in Puerto Rico. In addition to their loud calls, they compete with native birds and reptiles for insects. The mayor of Hilo, Hawaii, even declared a coquí state of emergency.

FOUL FACT

How do tiny frogs make such big sounds? With their nostrils closed, air from their lungs is pushed over their vocal chords and through their windpipe (trachea) into the air sac, the bulge beneath their chin. The air sac acts like a resonance chamber, a hollow space that amplifies the sound.

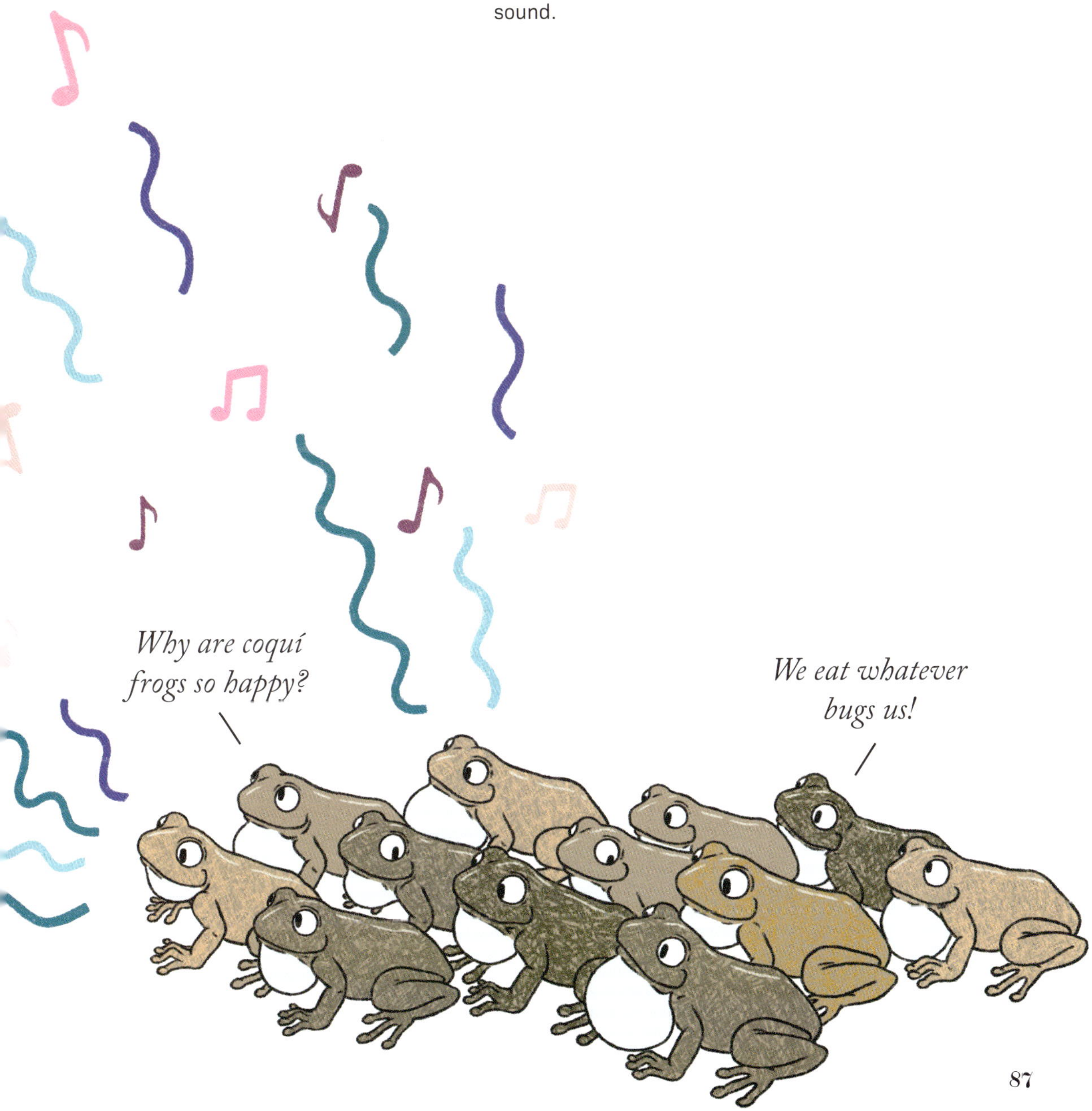
Why are coquí frogs so happy?
We eat whatever bugs us!

Frog Scientist

Cori Richards-Zawacki, Amphibian Biologist

How did you become interested in studying frogs?

I enjoyed science and math as a kid. In high school, I was part of a Science Olympiad program that introduced me to amphibians—animals including frogs, toads, newts, and salamanders that live both in water and on land. I also became interested in listening to frog calls.

***How ribbiting!* Tell us more!**

I grew up in Michigan and I joined a group of volunteers through a citizens' science program that seeks to answer questions like: How diverse is the local population of frogs and toads? Where can we find rare or invasive species? Are there long-term shifts in species diversity, range, and seasonal timing?

My mom would drive me to different sites, and we would record data about which frog calls we heard. One night, I recorded a species that was threatened, and no one had ever surveyed that species in that pond before. That was cool!

How do small frogs make such loud sounds?

Frogs have vocal cords like we do, but they also have a vocal sac that can amplify the vibrations from the vocal cords, which is what makes them so loud.

Why are frogs so loud?

Most of the time, you're hearing males making an "advertisement call." Their function is to attract females. The louder the frog, the better the chances a female will hear the call.

Why are they so noisy at night?

Many frogs are nocturnal, so they are more active and do most of their calling and mating at night. Some frogs (like many poison dart frogs) are diurnal and most active during the day.

Wouldn't being loud expose frogs to predators?

Yes! In fact, some Panamanian bats have learned to cue in on the exact elements of a male's call that is attractive to females, which makes those calls even more dangerous for a male. This has been well studied for the Túngara, a little brown frog that calls while floating on water in Central America.

What are some other "rude" frog behaviors?

The gastric brooding frog, a species that grows its babies in its stomach and then vomits them alive, is now extinct (no longer in existence). They were probably driven to extinction because of a fungus that's been attacking amphibians. Scientists were fascinated by these frogs because of their ability to turn off their gastric juices so they don't digest their own froglets.

What are you studying now?

I'm researching a species of poison dart frogs called *Oophaga pumilio*, which live on Panamanian islands. They get their toxicity (poison) from their specialized diet of ants and mites, which eat certain plants. We're trying to determine why frogs on different islands have different colors.

Do you have a pet frog?

A flower shop once called me to remove a Cuban tree frog that was hopping all over the store. It had apparently taken a ride on a plant they imported from Florida. After I caught it, my family and I kept the frog as a pet. We named it Valentino because I brought it home on Valentine's Day. It ate crickets.

Rude or Remarkable?

Now that you've come face-to-face with belchers, spitters, public poopers, pukers, gas passers, and more, let's show respect for these animals with attitude. Maybe they're not so rude after all. Which is your favorite?

FART!

Glossary

Amphibian An animal, such as a frog or salamander, that is able to live both on land and in water

Apex predator A predator (an animals that preys on others) at the top of the food chain, without natural predators of its own, such as a lion or a shark

Bioluminescence The production and emission of light by a living organism

Carrion The decaying flesh of dead animals

Conservation Preservation and protection of animals and their environment

Crustacean Spineless (invertebrate) animal with a hard exoskeleton, segmented body, and jointed appendages, such as crayfish, shrimp, and lobster

Decibel The unit that measures the intensity of sound

Ecology The study of the relationship between a living thing and its ecosystem

Ecosystem A community of interacting organisms

Extinct No longer in existence

Flatulence The ejection of gas produced during digestion in the stomach or intestines

Greenhouse gas Gas, such as carbon dioxide or methane, in the Earth's atmosphere that traps heat

Herpetologist A scientist who studies reptiles and amphibians

Krill Small invertebrates that resemble tiny shrimp and swim in huge swarms

Mucus A sticky, slimy substance made of water, proteins, sugar, and salt

Nematocyst A stinging capsule

Ornithologist A scientist who studies birds

Parasite An organism that lives on or in another organism (its host) and gets food at the expense of its host

Reptile An animal (such as a snake, lizard, turtle, or alligator) that is cold-blooded, lays eggs, and has a body covered with dry, scaly skin

Ruminant An animal, such as a cow or a sheep, with a stomach that has four compartments, allowing it to digest grass and other plants

Saliva A watery fluid produced by glands in the mouth that aids in digestion

Scavenger An animal that feeds on the remains of other animals

Symbiosis A beneficial relationship between two species

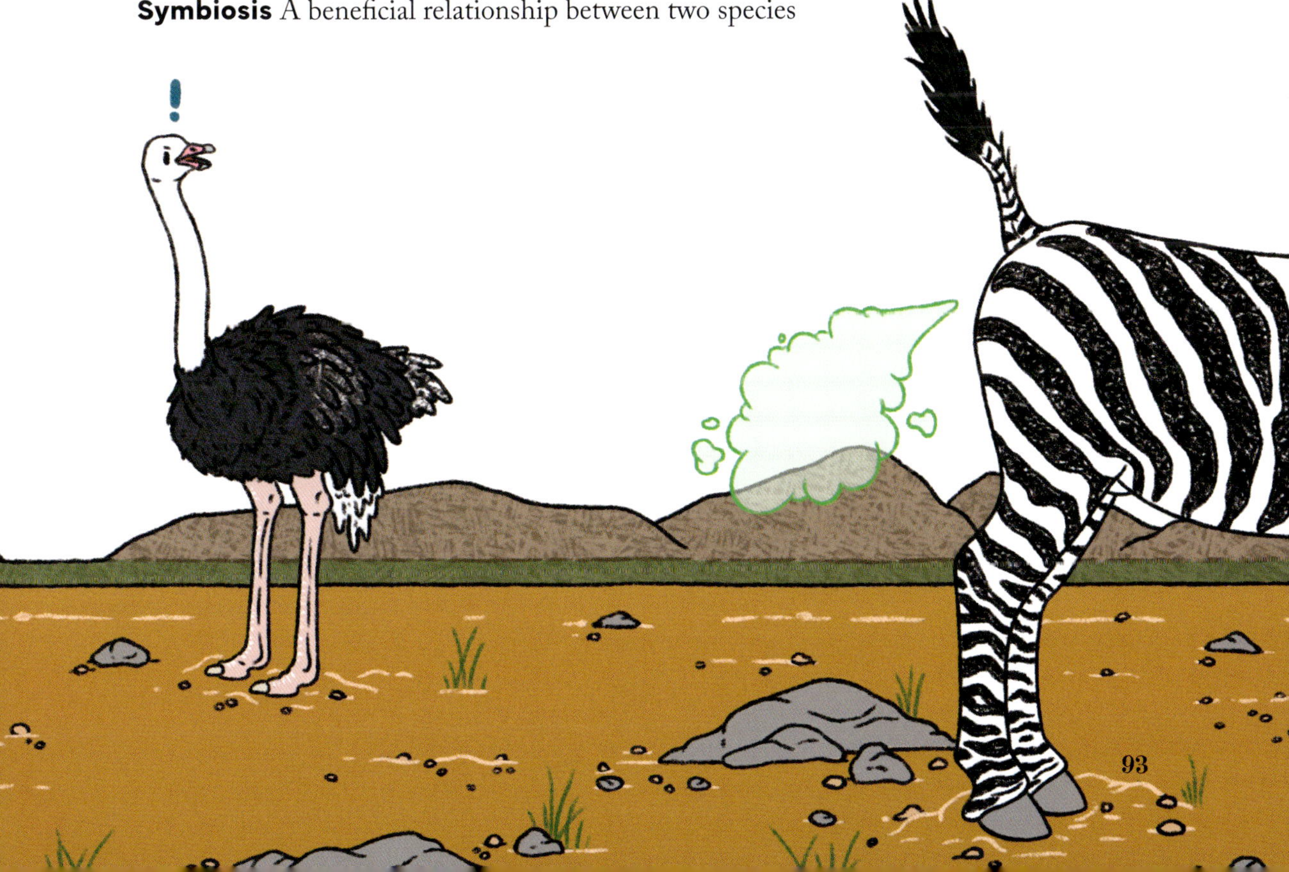

Index

About the Creators

Chana Stiefel is the award-winning author of more than thirty children's books, including *Animal Zombies & Other Real-Life Monsters*; *Let's Fly! Barrington Irving's Record-Breaking Flight Around the World*; and *Bravo, Avocado*. She grew up in South Florida, where she loved going on "swamp tromps" in the Everglades to search for snakes, lizards, frogs, gators, and gulping pelicans. Now she lives within spitting distance of New York City, where she enjoys sharing her passion for reading and writing with children. Learn more at chanastiefel.com.

Anna Louise Oliver is an artist who loves observing and drawing animals. She spent most of her childhood outside, climbing trees, looking for animals, and getting chased by the pigs that lived in the field across the road from her house. She studied illustration at the University of Plymouth and currently lives in southwest England, where she can normally be found drawing, watching movies, or getting excited about a dog she has just seen. This is the first book she has illustrated. You can see more of her work at annalouiseillustration.co.uk.

For my friends and family, rude dudes with attitude
—C. S.

For Izzy—my favorite rude animal
—A. L. O.

UNION SQUARE KIDS and the distinctive Union Square Kids logo are trademarks of Hachette Book Group, Inc.

ISBN 978-1-4549-5644-0 (hardcover)
ISBN 978-1-4549-5645-7 (paperback)
ISBN 978-1-4549-5646-4 (ebook)

Library of Congress Control Number: 2024049601

Union Square Kids books may be purchased in bulk for business, educational, or promotional use. For more information, please contact your local bookseller or the Hachette Book Group's Special Markets department at special.markets@hbgusa.com.

Printed in China

Lot #:

2 4 6 8 10 9 7 5 3 1

07/25

unionsquareandco.com

Cover and interior design by Marcie Lawrence
Cover and interior art by Anna Louise Oliver

Photos: 15 (left and top right): Antonio Cerullo's research snails; 15 (lower right): An upside-down jellyfish; 27 (all): Bombardier beetles; 38: Taline has a centralian carpet python around her neck; 39 (top): Taline's pet Sumatran short-tailed python, named Fawkes; 39 (middle): Special gloves are essential when handling poisonous snakes, like this baby Egyptian cobra; 39 (bottom): Taline's pet Jasper, a Centralian carpet python; 49 (all): Komodo dragons; 59 (all): Chinstrap penguins; 67 (top): Wombat poop cubes, part of a field study on Maria Island, Australia; 67 (bottom): A cassowary; 79 (middle): Yakira swims with rays; 79 (bottom): Atlantic Blacktip Shark; 89 (top): A strawberry poison dart frog; p. 89 (bottom): Cori finds frog eggs on a leaf in central Panama.

Getty Images: *E+:* Andworks: 67 (bottom); *iStock/Getty Images Plus:* LauraDin: 15 (bottom right); RibeirodosSantos: 49 (left); Marek Stefunko: 89 (right); USO: 49 (right, x2); Photos courtesy of the following: **Tom Altany:** 88; **Brookhaven National Lab:** 26; **Antonio Cerullo:** 14–15 (left and top right); **Brian Gratwicke:** 89 (left); **Taline Kazandjian:** 38–39; **Moorea Moana Tours:** 79 (middle); **© National Academy of Sciences, U.S.A., 1999:** 27; **Secrets in the Scat:** 66, 67 (top); **Shark Research and Conservation Lab at the University of Miami:** 79 (bottom); **Smithsonian's National Zoo and Conservation Biology Institute:** Meghan Murphy: 48; **Abbie Sophia:** 79 (top); **Noah Strycker:** 58–59